Ergodic Theory and Related Topics

Mathematical Research · Mathematische Forschung

Wissenschaftliche Beiträge
herausgegeben von der
Akademie der Wissenschaften der DDR
Institut für Mathematik

Band 12

Ergodic Theory and Related Topics

Ergodic Theory and Related Topics

Proceedings of the Conference held in Vitte/Hiddensee (GDR), October 19–23, 1981

edited by Prof. Dr. Horst Michel
Martin-Luther-Universität Halle-Wittenberg

Akademie-Verlag · Berlin 1982

Die Titel dieser Schriftenreihe werden vom Originalmanuskript der Autoren reproduziert.

Erschienen im Akademie-Verlag, DDR-1086 Berlin, Leipziger Str. 3—4
Lektor: Dr. Reinhard Höppner

Lizenznummer: 202 · 100/422/82
Gesamtherstellung: VEB Kongreß- und Werbedruck, 9273 Oberlungwitz
Umschlag: Rolf Kunze
Bestellnummer: 763 138 7 (2182/12) · LSV 1075
Printed in GDR
DDR 30,— M

INTRODUCTION

During the week of October 19-23, 1981, a conference on "Ergodic Theory and Related Topics" was held in Vitte (Hiddensee), GDR. In this conference specialists from Austria, Czechoslovakia, Federal Republic of Germany, France, German Democratic Republic, Great Britain, Netherlands, Poland and USSR participated.

It was the aim of the meeting to represent recent development in measure-theoretic and topological dynamical systems as well as connections to fields, which are in the neighbourhood of ergodic theory. This volume contains most of the material of the conference and some additional papers.

The organizer would like to thank in particular each of the participants for contributing to the success of the conference, the Mathematical Department of the Martin-Luther-University Halle-Wittenberg and the Mathematical Society of the German Democratic Republic for sponsoring this conference.

Especially Prof. Dr. O. Krötenheerdt, Dr. H.-H. Buchsteiner, Dr. K. Häsler, T. Jentsch, B. Mathiszik, G. Schwenzfeger, V. Warstat and Dr. E. Weihert from the Martin-Luther-University, Prof. Dr. R. Klötzler and I. Bausch from the Mathematical Society of the German Democratic Republic and Dr. R. Höppner from the Akademie-Verlag Berlin offered their help and advice in organizing this conference and in preparing these Proceedings.

Halle (Saale), March 1982 Horst Michel

CONTENTS

LIST OF PARTICIPANTS OF THE CONFERENCE

Dr. sc. Alberti, P.M.	GDR
Böhme, O.	GDR
Dr. habil. Bothe, H.-G.	GDR
Dr. Brudno, A.	USSR
Prof. Dr. Bunimovich, L.	USSR
Dr. Crell, B.	GDR
Prof. Dr. Denker, M.	FRG
Dr. Der, R.	GDR
Dr. Dravecký, J.	Czechoslovakia
Dr. Englisch, H.	GDR
Franke, M.	GDR
Dr. Freudenberg, W.	GDR
Dr. Häsler, K.	GDR
Prof. Dr. Helmberg, G.	Austria
Doz. Dr. Hofbauer, F.	Austria
Prof. Dr. Jacobs, K.	FRG
Jentsch, Th.	GDR
Dr. Jetschke, G.	GDR
Dr. del Junco, A.	Netherlands
Doz. Dr. Kamiński, B.	Poland
Prof. Dr. Keane, M.	Netherlands
Dr. Keller, G.	FRG
Dr. Komorník, J.	Czechoslovakia
Prof. Dr. Krengel, U.	FRG
Doz. Dr. Kwiatkowski, J.	Poland
Manthey, R.	GDR
Mathiszik, B.	GDR
Prof. Dr. Michel, H.	GDR
Dr. Naumann, W.	GDR

Prof. Dr. Neubrunn, T.	Czechoslovakia
Prof. Dr. Parry, W.	Great Britain
Dr. Pesin, J.	USSR
Doz. Dr. Riečan, B.	Czechoslovakia
Prof. Dr. Schmidt, K.	Great Britain
Schwenzfeger, G.	GDR
Dr. Siegmund-Schultze, R.	GDR
Dr. Stelzer, B.	GDR
Dr. sc. Šujan, Št.	Czechoslovakia
Prof. Dr. Szlenk, W.	Poland
Prof. Dr. Thouvenot, J.-P.	France
Prof. Dr. Uhlmann, A.	GDR
Warstatt, V.	GDR
Dr. Weihert, E.	GDR
Dr. Wobst, R.	GDR
Zylka, Ch.	GDR

LIST OF TALKS OF THE CONFERENCE

Bothe,H.-G.:	Structure and position of attractors.
Brudno,A.:	Entropy and algorithmic complexity in dynamical systems, control theory and information theory.
Bunimovich,L.:	Some rigorous results in nonequilibrium statistical mechanics.
Denker,M.:	Statistical decision procedures and Ergodic Theory.
Englisch,H.:	The spectrum of Schrödinger operators with random potential.
Freudenberg,W.:	A weak ergodic theorem for nonlinear averaging procedures.
Häsler,K.:	Lower and upper bounds of dynamical systems.
Helmberg,G.:	Entropy of simple broken linear functions.
Hofbauer,F.:	Piecewise monotonic transformation of an interval.
Jacobs,K.:	Recent developments in Ergodic Theory.
Jetschke,G.:	Random fields and stochastic partial differential equations related to physics.
del Junco,A.:	Prime transformations in Ergodic Theory and Topological Dynamics.
Kamiński,B.:	The theory of invariant partitions for Z^d-actions.
Keane,M.:	Finitary isomorphisms.
Keller,G.:	Piecewise expanding transformations.
Komorník,J.:	Theorems of Goodwyn's type.
Krengel,U.:	Multiparameter subadditive ergodic theorems.
Kwiatkowski,J.:	Metric and spectral isomorphisms of Morse dynamical systems.
Michel,H.:	Set configuration spaces.
Neubrunn,T.:	Compatibility in Quantum Logics.
Parry,W.:	On the classification of finite state Markov chains.
Pesin,J.:	Gibbs measures for partial-hyperbolic dynamical attractors.
Riečan,B.:	On the individual ergodic theorem.

Schmidt,K.:	Two applications of a theorem by Glimm and Effros in Ergodic Theory.
Siegmund-Schulze,R.:	Ergodic properties of infinitely divisible point processes.
Stelzer,B.:	λ-continuous Markov chains and laws of large numbers.
Šujan,Št.:	Codes in ergodic theory and information: some examples.
Thouvenot,J.-P.:	Invariants.
Uhlmann,A.:	Stochastic maps in the dual of C^*- and W^*-algebras.
Weihert,E.:	Maximal embeddability of totally ergodic transformations with quasidiscrete spectrum in flows.

DISTRIBUTION OF INVARIANT MEASURES ON EXPANDING ATTRACTORS IN THE AMBIENT MANIFOLD

H.G.Bothe[1)]

1. Introduction Let M be an m-dimensional smooth manifold without boundary and $f:M \longrightarrow M$ a C^2-diffeomorphism with a hyperbolic attractor Λ (precise definitions are given in section 2). Then Ruelle [6] has constructed an f-invariant normed Borel measure on Λ for which - roughly speaking - $\mu_\Lambda(A)$ (A a Borel subset of Λ) is the ratio of time for which the foreward orbit $(f^i(p))$ $(i = 0,1,2,\ldots)$ of a generic point p in the basin of Λ stays near A. We are interested in the relations between the geometric structure of Λ and properties of μ_Λ. Especially we ask how μ_Λ is distributed in M. For the questions and results which will be discussed in this talk we need the following definition:

Let μ be any Borel measure on M and $d \geq 0$ a real number. We say that μ has no d-dimensional mass concentration if for each $p \in M$

$$\liminf_{\rho \to +0} \frac{\log \mu(B(p,\rho))}{\log \rho} > d$$

holds where $B(p,\rho)$ denotes the geodesic ball with centre p and radius ρ with respect to an arbitrarily chosen Riemannian metric of M. This means that for $\rho \to +0$ the measure of $B(p,\rho)$ goes to 0 quicker than ρ^d. We define $d(\mu)$ to be the supremum of all d for which μ has no d-dimensional mass concentration. It seems reasonable to say that $d(\mu) > d(\mu')$ implies that μ is more smoothly distributed in M than μ'.

There are simple examples which show that $d(\mu)$ is not determined by the support $\operatorname{supp}\mu$ of μ. But in special cases there are relations between the Hausdorff dimension $\dim_H \Lambda$ of a hyperbolic attractor Λ and $d(\mu_\Lambda)$, and in a subsequent paper we shall ask for conditions under which $d(\mu_\Lambda) = \dim_H \Lambda$ holds. In this talk we consider 1-dimensional hyperbolic attractors Λ and show that there is a positive real number ε depending in a simple way on the dynamic of f near Λ such that $d(\mu_\Lambda) > 1 + \varepsilon$ holds. The main tool are tubular neighborhoods of Λ in M as introduced in [1] or [2] and

1) Institut für Mathematik der Akademie der Wissenschaften der DDR
DDR 1080 Berlin, Mohrenstr. 39

R. Williams' representation of Λ by an expanding mapping $\varphi:\Sigma\to\Sigma$ of a 1-dimensional branched manifold Σ [7], [8]. The main point is that associated with Σ there is a normed φ-invariant Borel measure μ_Σ on Σ which completely describes μ_Λ.

2. Definitions and some basic facts Let $f:M\to M$ be a C^r diffeomorphism $(r\geq 2)$ of an m-dimensional C^∞ manifold M without boundary and Λ a compact invariant subset of M (i.e. $f(\Lambda)=\Lambda$). We call Λ hyperbolic if there are real numbers a, λ satisfying $0<a$, $0<\lambda<1$ and C^0 subbundles $T^s(\Lambda)$, $T^u(\Lambda)$ of the restriction $T_\Lambda(M)$ of the tangent bundle $T(M)$ of M to Λ such that

$$T_\Lambda(M) = T^s(\Lambda)\oplus T^u(\Lambda)$$

and with respect to a Riemannian metric on M

$$|df^i(v)|\leq a\cdot\lambda^i|v|,\qquad |df^{-i}(w)|\leq a\cdot\lambda^i|w|$$

holds for all $v\in T^s(\Lambda)$, $w\in T^u(\Lambda)$ and $i=0,1,2,\ldots$. If Λ is hyperbolic, then for each $p\in\Lambda$ there are one-to-one C^r immersions

$$w_p^s:\mathbb{R}^{n'}\to M,\qquad w_p^u:\mathbb{R}^n\to M$$

such that $w_p^s(o)=w_p^u(o)=p$,

$$w_p^s(\mathbb{R}^{n'}) = W_p^s = \{q\in M;\ \lim_{i\to\infty}\operatorname{dist}(f^i(q),f^i(p)=0\},$$

$$w_p^u(\mathbb{R}^{n}) = W_p^u = \{q\in M;\ \lim_{i\to-\infty}\operatorname{dist}(f^i(q),f^i(p)=0\},$$

and

$$T_p(W_p^s) = T_p^s(\Lambda),\quad T_p(W_p^u) = T_p^u(\Lambda)$$

holds. These sets W_p^s, W_p^u are called the stable and the unstable manifold of p.

We call a compact hyperbolic invariant set Λ attracting if it has a neighborhood U in M satisfying $f(U)\subset U$, $\bigcap_{i=0}^{\infty} f^i(U)=\Lambda$. The basin $B(\Lambda)=\bigcup_{p\in\Lambda}W_p^s$ of an attracting set Λ is the set of all points $q\in M$ which are attracted by Λ in the sense that $\lim_{i\to\infty}\operatorname{dist}(f^i(q),\Lambda)=0$. $B(\Lambda)$ is an open neighborhood of Λ in M. It is not hard to see that Λ is attracting iff for each $p\in\Lambda$ the whole unstable manifold W_p^u lies in Λ. Each attracting hyperbolic set Λ contains a finite number of minimal attracting subsets $\Lambda_1,\ldots,\Lambda_r$ (i.e. each Λ_i is compact, invariant and attracting but does not contain any proper subset which is also compact, invariant and attracting). Since $B(\Lambda_1)\cup\ldots\cup B(\Lambda_r)$ is a dense and open subset of $B(\Lambda)$, these sets Λ_i can be regarded as the essential parts of Λ, and we shall consider only minimal hyperbolic attracting sets which will be called hyperbolic attractors.

The hyperbolic attractors whose structure is known best are the expanding attractors Λ which are characterized by the fact that their topological dimension coincides with the dimension of their unstable manifolds W_p^u $(p \in \Lambda)$ or, equivalently, that each component of a sufficiently small open subset of Λ is contained in an unstable manifold W_p^u of Λ.

In this paper we consider 1-dimensional hyperbolic attractors Λ. Moreover we assume that Λ is connected. This is an inessential restriction, since a hyperbolic attractor has a finite number, say r, of components which are attractors of f^r. The theorem below holds for 1-dimensional hyperbolic attractors, but a probably possible generalization to higher dimensional attractors, if it can be obtained by similar methods, will be restricted to expanding attractors. (All 1-dimensional hyperbolic attractors are expanding.)

Let Λ be a (connected) 1-dimensioal hyperbolic attractor. Then the family of all stable manifolds W_p^s $(p \in \Lambda)$ is a C^1 foliation $\mathfrak{W}^s$ of $B(\Lambda)$ whose leaves are $(m-1)$-dimensional. Moreover, by [2] Λ has a neighborhood N in $B(\Lambda)$ (called tubular neighborhood) which has the following properties:

(1) N is a compact connected m-manifold.

(2) $f(N) \subset \mathrm{Int}(N)$

(3) There are finitely many critical points of $\mathrm{Bd}(N)$, i.e. points at which $\mathrm{Bd}(N)$ is tangent to the corresponding leaf of $\mathfrak{W}^s$.

(4) For each $p \in \Lambda$ the component of $W_p^s \cap N$ containing p is an $(m-1)$-disk or the union of two $(m-1)$-disks with one common boundary point.

(5) For each $p \in \Lambda$ each component of $W_p^s \cap N$ intersects Λ.

If we identify points in N which lie for a leaf $W^s \in \mathfrak{W}^s$ in the same component of $N \cap W^s$, we get a compact space Σ which is called a branched 1-manifold (see [7] or [8]), i.e. Σ may be represented by a compact subset of $\mathbb{R}^3$ which can be covered by a finite family $\mathfrak{A}$ of smooth arcs such that the following conditions are satisfied:

(1) If $A, A' \in \mathfrak{A}$, then $A \cap A'$ contains at most one point, and this point is end point of at least one of the arcs A, A'.

(2) Each $x \in \Sigma$ lies in at most two arcs of $\mathfrak{A}$.

(3) If $x \in \Sigma$ lies in two arcs of $\mathfrak{A}$, then the tangents at x of these arcs coincide.

(4) Each $x \in \Sigma$ lies in the interior of an arc which is contained in Σ (i.e. Σ has no end points).

A point $x \in \mathfrak{A}$ which is end point of exactly one arc in $\mathfrak{A}$ is called a branch point of Σ. The C^1 mapping $\varphi : \Sigma \to \Sigma$ which is defined by the commutative diagram

$$\begin{array}{ccc} N & \xrightarrow{f} & N \\ \pi\downarrow & & \downarrow\pi \\ \Sigma & \xrightarrow{\varphi} & \Sigma \end{array} \qquad (*)$$

with $\pi : N \to \Sigma$ the natural projection is expanding is the sense that there are real numbers $a > 0$, $\lambda > 1$ such that

$$|d\varphi^i(v)| \geq a \cdot \lambda^i |v| \qquad (i = 0,1,2,\ldots) \qquad (**)$$

holds for all tangent vectors v of Σ where $|v|$ denotes the length of the vector v in $\mathbb{R}^3$. The representation of Σ in $\mathbb{R}^3$ can be chosen so that $a = 1$, and we omit this factor a in (**). The attractor Λ and the restriction of f to Λ are uniquely determined by Σ and $\varphi : \Sigma \to \Sigma$. Indeed, Williams proved that Λ can be identified with the inverse limit of the sequence $\Sigma \xleftarrow{\varphi} \Sigma \xleftarrow{\varphi} \cdots$ and that f on Λ is the shift mapping, i.e. if $p \in \Lambda$ corresponds to $(x_0, x_1, \ldots)$ in the inverse limit, then $f(p)$ corresponds to $(\varphi(x_0), \varphi(x_1), \ldots) = (\varphi(x_0), x_0, x_1, \ldots)$. This is the reason why we call $\varphi : \Sigma \to \Sigma$ or the diagram (*) a W-representation of Λ.

As shown by D. Ruelle (see [6], [3], or [5]) each hyperbolic attractor Λ carries an f-invariant normed Borel measure μ_Λ which is determined by each of the following properties:

(A) There is a subset A of $B(\Lambda)$ which is a zero set (with respect to the differential structure of M) such that for each $p \in B(\Lambda) \setminus A$ and for each continuous function $\tau : M \to \mathbb{R}$ we have

$$\lim_{i \to \infty} \frac{1}{i} \sum_{k=1}^{i} \tau(f^k(p)) = \int \tau \, d\mu_\Lambda.$$

(B) If μ is a normed measure on $B(\Lambda)$ which is given by a continuous density function (with respect to the Lebesgue measure associated with a Riemannian metric of M), then μ_Λ is the vague limit of $f^i \mu$ $(i = 0,1,2,\ldots)$.

3. The measure μ_Σ In the following simple description of μ_Λ for 1-dimensional hyperbolic attractors we use additively continuous density functions on a branched 1-manifold Σ which are defined as follows: A function $\sigma : \Sigma \to \mathbb{R}$ is called additively continuous if σ is continuous at each non branch point of Σ and if for each branch point $x \in \Lambda$ we have

$$\sigma(x) = \lim_{y_1 \to x} \sigma(y_1) + \lim_{y_2 \to x} \sigma(y_2),$$

where y_1 and y_2 converge to x on different branches belonging to x as indicated in the following figure:

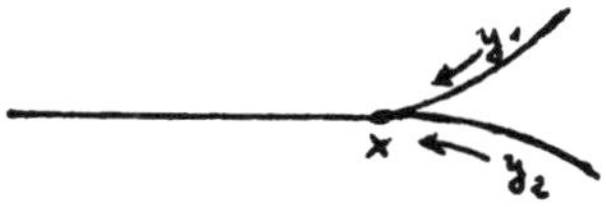

Proposition. There is a φ-invariant normed Borel measure μ_Σ on Σ which is given by a positive, additively continuous density function $\sigma : \Sigma \to \mathbb{R}$ such that $\pi : \Lambda \to \Sigma$ is measure preserving with respect to the measures μ_Λ and μ_Σ.

Remark. Since the family of all subsets $f^i \pi^{-1}(\Xi)$ $(i = 0,1,\dots;$ Ξ closed in Σ) is a basis for the Borel sets in Λ, the proposition shows that μ_Λ is uniquely determined by μ_Σ.

Proof: Since μ_Λ does not change if we replace f by f^i $(i \geq 1)$ and since φ has periodic points (see [7] or [8]), we may assume that φ has a fixed point x_0. It is not hard to find a C^1 mapping $\xi : [0, \infty) \to \Sigma$ and an increasing C^1 mapping $\tau : [0, \infty) \to [0, \infty)$ such that $\xi(0) = x_0$, $\tau(0) = 0$, $\varphi\xi(t) = \xi\tau(t)$ $(t \geq 0)$. This mapping ξ is periodic (if $t > 0$, $\xi(t) = x_0$, then $\tau(t)$ is a period), and we denote by t_0 the minimal period of ξ. Then by

$$e^{2\pi it} \mapsto \xi(t_0 \cdot t)$$

we get a surjective C^1 immersion $\rho : S^1 \to \Sigma$ of the unit circle S^1 onto Σ. (This mapping is surjective since in a W-representation $\varphi : \Sigma \to \Sigma$ of a 1-dimensional expanding attractor for each subarc A of Λ there is an exponent i such that $\varphi^i(A) = \Sigma$.) If we define $\varphi^* : S^1 \to S^1$ by $\varphi^*(e^{2\pi it}) = e^{2\pi i \tau(t_0 \cdot t)/t_0}$, we get an expanding C^1 mapping of S^1 for which the following diagram is commutative

$$\begin{array}{ccc} S^1 & \xrightarrow{\varphi^*} & S^1 \\ \rho\downarrow & & \downarrow\rho \\ \Sigma & \xrightarrow{\varphi} & \Sigma \end{array}$$

By [4] there is a unique φ^*-invariant C^0 measure on S^1 whose density function will be denoted by σ^*. Then

$$\sigma(x) = \sum_{\rho(s)=x} \left|\frac{d\rho}{ds}(s)\right|^{-1} \cdot \sigma^*(s)$$

defines an additively continuous function σ on Σ which is the density function of a φ-invariant measure μ_Σ. To prove that π is measure preserving we consider a normalized C^0 measure μ on M whose support lies in N and for which $\pi(\mu) = \mu_\Sigma$ holds. Then for $i = 0,1,2,\ldots$ we have

$$\pi f^i(\mu) = \varphi^i(\mu_\Sigma) = \mu_\Sigma,$$

and the property (B) of μ_Λ implies that $f^i(\mu)$ converges to μ_Λ. This proves $\pi\mu_\Lambda = \mu_\Sigma$.

<u>4. A lower bound for $d(\mu_\Lambda)$</u> As above let Λ be a connected 1-dimensional hyperbolic attractor of $f:M \to M$. It is well known that M carries an adapted metric with respect to Λ i.e. a Riemannian metric for which positive real numbers $\alpha > 1$, $\alpha^* < 1$ exist such that

$$|df(v)| \geq \alpha |v|, \qquad |df(w)| \leq \alpha^* |w|$$

holds for all $v \in T^u(\Lambda)$, $w \in T^s(\Lambda)$. Let α be chosen maximal and denote by β the maximal number satisfying

$$|df(w)| \geq \beta |w|$$

for all $w \in T^s(\Lambda)$. Since Λ is compact, β is positive.

<u>Theorem.</u> For the invariant measure μ_Λ of a 1-dimensional hyperbolic attractor Λ

$$d(\mu_\Lambda) \geq 1 - \frac{\log \alpha}{\log \beta}$$

holds. (Since $\beta \leq \alpha^* < 1$, we have $-\frac{\log \alpha}{\log \beta} > 0$.)

Remark 1. Considering the volume of the images $f^i(N)$ (N a tubular neighborhood of Λ) we find $\alpha\cdot\beta \leq 1$ and therefore $-\frac{\log\alpha}{\log\beta} \leq 1$. This raises the question whether $d(\mu_\Lambda) \leq 2$ holds, and, indeed, using the Hausdorff dimension of Λ we can give a positive answer.

Remark 2. In our definition the numbers α, β depend on the Riemannian metric and are not necessarily optimal. To get invariant values for α, β one should define α to be the average of the function

$$\alpha(p) = |d_p f| = \inf\left(|d_p f(v)|;\ v \in T^u_p(\Lambda),\ |v| = 1\right)$$

and choose β in a similar way.

Proof. Let N be a tubular neighborhood of Λ and consider the corresponding W-representation

$$\begin{array}{ccc} N & \xrightarrow{f} & N \\ \pi\downarrow & & \downarrow\pi \\ \Sigma & \xrightarrow{\varphi} & \Sigma \end{array}$$

For each $\eta > 0$ we can choose a Riemannian metric for Σ such that there is a real number $\alpha' > \alpha - \eta$ satisfying

$$|d\varphi(v)| \geq \alpha' |v| \qquad (***)$$

for all tangent vectors v of Σ. An elementary argument shows that for each $\beta' < \beta$ there is a tubular neighborhood N of Λ such that

$$\operatorname{dist}(\Lambda, f^i(\operatorname{Bd}(N))) \geq \beta'^i \operatorname{dist}(\Lambda, \operatorname{Bd}(N)) \qquad (****)$$

holds for $i = 0,1,2,\ldots$. Therefore to prove the theorem it is sufficient to take a tubular neighborhood N of Λ and a Riemannian metric for Σ and to prove

$$\liminf_{\rho \to 0} \frac{\log \mu_\Lambda(B(p,\rho))}{\log \rho} \geq 1 - \frac{\log \alpha'}{\log \beta'}$$

where α', β' are positive real numbers satisfying $(***)$ or $(****)$ respectively.

As above the measure $\pi\mu_\Lambda$ will be denoted by μ_Σ. Since μ_Σ is defined by an additively continuous density function on Σ, we can find a real number $\gamma > 0$ satisfying

$$\mu_\Sigma(\Xi) < \gamma \cdot \operatorname{diam} \Xi$$

for each Borel subset Ξ of Σ. We denote $\operatorname{dist}(\Lambda, \operatorname{Bd}(N))$ by δ and choose $\xi > 0$ so that for $p \in \Lambda$, $x \in \Sigma$ and $\varepsilon > 0$ sufficiently small $\operatorname{dist}_\Sigma(x, \pi(p)) > \xi\varepsilon$ implies $\operatorname{dist}_\Lambda(p, \pi^{-1}(x)) > \varepsilon$.
For $p \in \Lambda$, $\rho < \beta'^i \delta$ we have $B(p,\rho) \subset f^i(N)$. Let $i(\rho)$ be the integer satisfying

$$\frac{\log\rho}{\log\beta'} - \frac{\log\delta}{\log\beta'} - 1 \le i(\rho) < \frac{\log\rho}{\log\beta'} - \frac{\log\delta}{\log\beta'}$$

(ρ sufficiently small). Then $B(p,\rho) \subset f^{i(\rho)}(N)$.

Denote by U the $(\xi\rho\alpha'^{-i(\rho)})$-neighborhood of $\pi f^{-i(\rho)}$ in Σ. Then by a simple geometric argument we have

$$B(p,\rho) \subset f^{i(\rho)}\pi^{-1}(U)$$

and

$$\mu_\Lambda(B(p,\rho)) \le \mu_\Lambda(f^{i(\rho)}\pi^{-1}(U)) = \mu_\Lambda(\pi^{-1}(U)) = \mu_\Sigma(U) \le 2\gamma\xi\rho\alpha'^{-i(\rho)}.$$

This implies

$$\log\mu_\Lambda(B(p,\rho)) \le \log 2\gamma + \log\xi + \log\rho - \log\alpha'\left(\frac{\log\rho}{\log\beta'} - \frac{\log\delta}{\log\beta} - 1\right)$$

and

$$\frac{\log\mu_\Lambda(B(p,\rho))}{\log\rho} \ge 1 - \frac{\log\alpha'}{\log\beta'} + \frac{1}{\log\rho}\left(-\log 2\gamma + \log\xi + \frac{\log\alpha'\log\delta}{\log\beta'} + \log\alpha'\right)$$

Now

$$\liminf_{\rho\to 0} \frac{\log\mu_\Lambda(B(p,\rho))}{\log\rho} \ge 1 - \frac{\log\alpha'}{\log\beta'}$$

follows immediately.

References

[1] H.G.Bothe: The ambient structure of 1-dimensional hyperbolic attractors in 3-manifolds. Report Inst.Math. Akad.Wiss. DDR, 01/1981.

[2] H.G.Bothe: The ambient structure of expanding attractors. I.Local triviality and tubular neighborhoods. (to appear).

[3] R.Bowen: Equilibrium states and ergodic theory of Anosov diffeomorphisms. Springer Lecture Notes in Math. 470 (1975).

[4] K.Krzyzewski: Some results on expanding mappings. Dynamical Systems III - Warsaw, Astérisque 50 (1978), 205-218.

[5] S.E.Newhouse: Lectures on dynamical systems. Dynamical Systems CIME Lectures 1978, Birkhäuser Progress in Math. 8 (1980), 1-114.

[6] D.Ruelle: A measure associated with Axiom A attractors. Amer. J. Math. 98 (1976), 619-654.

[7] R.Williams: One dimensional non-wandering sets. Topology 10 (1971), 133-147.

[8] R.Williams: Expanding attractors. Publ. Math. IHES 43 (1974), 169-203.

DYNAMIC SYSTEMS AND ALGORITHMIC THEORY

A.Brudno[1)]

0.Introduction.

The report considers an algorithmic approach to dynamic systems theory. Methods of algorithmic theory and symbolic dynamics are applied to assign every dynamic system trajectory a number - the minimum quantity of information necessary for its restoration, which characterises the algorithmic complexity of its behaviour in the phase space. Correspondence between trajectory's complexity and entropy characteristics , universal coding problem of information theory, dynamic systems simulation problems is established.

Support and encouraging, constantly received by the author from V.M. Alexeyev can scarcely be exaggerated.

1. Definitions.

1.1. Kolmogorov complexity of finite objects [1] .

Let A be a computable function (algorithm), defined on a set of words of alphabet $\{0,1\}$, receiving values on a set of words of some alphabet L. Denote $\ell(p)$ the length of word p (i.e. the number of symbols it contains).

Complexity of word s relative to algorithm A is defined by equality:

$$K_A(s) \triangleq \min \ell(p) \qquad | \ A(p) = s$$

There exists an asymptotically optimal algorithm B , when inequality $K_B(s) \leq K_A(s) + C_A$ holds for every word s and any algorithm A (C_A denotes a constant, dependent only on A).
By definition $K(s) \triangleq K_B(s)$.

1.2. Complexity of dynamic system trajectory [2] .

Let $u = \{U_i\}_{i=1}^{M}$ be a cover of space X . The symbolic dynamic of system (X, T) corresponding to the cover u , is defined as a mapping

$$\varphi_u : X \rightarrow \Omega_u .$$

where

$$\omega = \{\omega(i)\}_{i=0}^{\infty} \in \varphi_u(x) \Longleftrightarrow T^i x \in U_{\omega(i)} \quad \text{for all } i.$$

---

1) Institute of Systems Studies, Ryleeva 29 , 119034 Moscow , USSR

Complexity of a point x trajectory relative a cover u is by definition:

$$K(x,T|u) \overset{\Delta}{=} \overline{\lim_{n\to\infty}} (1/n) \min_{\omega \in \varphi_u(x)} K(\omega^n),$$

where $\omega^n \overset{\Delta}{=} \omega(0)\omega(1)\ldots\omega(n-1)$.

By the complexity of point X trajectory in a topological dynamic system (X,T) we mean

$$K(x,T) \overset{\Delta}{=} \sup_u K(x,T|u) \qquad | u \text{ - open finite cover.}$$

Properties of quantities $K(x,T|u)$, $K(x,T)$ are in detail considered in [3]. As it follows are the main results concerning ergodic theory, information theory and dynamic system simulation problems.

2. Ergodic theory.

The algorithmic complexity of a trajectory is linked with the measure and topological entropy by the following propositions:

Proposition 2.1. Let measure μ -ergodic, u -partition, then

$$K(x,T|u) = h_\mu(T,u) \qquad \text{for } \mu\text{-almost every } x \in X.$$

Proposition 2.2. For ergodic measure μ

$$K(x,T) = h_\mu(T) \qquad \text{for } \mu\text{-almost every } x \in X.$$

Proposition 2.3. For all $x \in X$

$$K(x,T) \leqslant h(T),$$

and there exist dynamic systems with strict equality for all $x \in X$.

A more precise upper bound for a trajectory's complexity is given by entropy according to Kamae [4].

Proposition 2.4. Let $V_T(x)$ - a set of weak limited measures for point x, $h_T(x) \overset{\Delta}{=} \sup_\mu h_\mu^T(T)$ its Kamae's entropy.

Then for all $x \in X$

$$K(x,T) \leqslant h_T(x).$$

3. Information theory.

The problem of efficient coding of dynamic system trajectories is closely linked with the problem of transmission through a discrete noiseless channel of an ensemble of messages with an unknown statistic.

Investigating the properties of messages algorithmic complexity enables to receive a series of results concerning this problem.

Proposition 3.1. There exists an effective universal coding providing the limit for the average length of code words equal the entropy not only with Pr. 1 ([5], [6]), but almost everywhere.

Proposition 3.2. The channel capacity (topological entropy for symbolic dynamic system (Ω, σ) is given by

$$C(\Omega) = \lim_{n\to\infty} \sup_{\omega\in\Omega} K(\omega^n)/n$$

Proposition 3.3. The message's algorithmic complexity $K(\omega) \overset{\Delta}{=} \overline{\lim_{n\to\infty}} K(\omega^n)/n$ appears to be the best in the class of characteristics coinciding with the entropy on the set of full measure.

Proposition 3.4. The equation $K(\omega) = h_\mu(\sigma)$ gives a sufficient but not necessary condition for the randomness of ω in the sense of Martin-Löff [7].

4. Dynamic system simulation.

In terms of an algorithmic approach to dynamic system theory it became possible to answer the following question: how can it happen that while simulating (real or numerical) strongly determinated systems there appear trajectories with a random behaviour, i.e. irregular under any law. This fact was recently observed by many authors while investigating dynamic systems of mathematical, physical and applied origin.

In order to explain the phenomena consider the process of measuring in a "symbolic interpretation", belonging to Alexeyev [8].

In reality observing a system trajectory means registrating indices of some device which on behalf of its limited measurement capacity receives only a finite number of values $1,2,\dots, M$, e.g. $X = \bigcup_{i=1}^{M} E_i$. Indeed when point X falls into region E_i, (at the moment $k: T^k_X \in E_i$) the device shows i. The sequence of indices $\omega = \{\omega(i)\}_{i=0}$ represents the experimental data on dynamic system (X,T). It may be treated stochastically by methods of algorithmic theory and on this ground (by a sound choice of the device) come to an appropriate conclusion on the behaviour of the original trajectory.

Within the bounds of algorithmic theory a truly random sequence is recognized, roughly speaking, by the complexity of its fragments - $K(\omega^n)$ - being proportional to their length.

It appears [7], that sequences random in the above described sense satisfy all the statistical tests, which provides an additional argument in favor of this generally accepted definition of randomness . Consequently, holds

Proposition 4.1. Random trajectories are obligatory to appear and even be typical when simulating (real or numerical) dynamic systems of non-zero entropy.

It means that statistical analysis of experimental data cannot reveal the deterministic or stochastic origin of the process under consideration.

References

[1] Kolmogorov A.N., "Three Approaches to the Definition of the "Amount of Information Concept", Problems in Information Transmission, v.1(1965), 1, 3-7.

[2] Brudno A.A., "On the complexity of Dynamic System Trajectory", UMN, 33(1978), 1, 207-208.

[3] Brudno A.A., "Entropy and Algorithmic Complexity of Dynamic Systems Trajectory", Preprint, M, The Institute for Systems Studies, 1980; Procedings of Moscow Mathematical Society, v.144 (to appear).

[4] Kamae T., "Normal Numbers and Ergodic Theory", Lecture Notes in Math., v.550, Springer-Verlag, 1976, 253-269.

[5] Fitingof B., "Optimal Coding in the case of unknown and changing message statistics", Problems in Information Transmission, v.2(1966), 2, 3-11.

[6] Davisson L., "Universal Noiseless Coding", IEEE Trans. Information Theory, IT-19(1973), 6, 783-795.

[7] Martin-Löff P., "The Definition of Random Sequences", Information and Control, 9(1966), 619-682.

[8] Alexeyev V.M., "Symbolic Dynamics, The Eleventh Mathematical School", Kiev, Publishing House of the Institute of Mathematics of the Ukrainian Academy of Sciences, 1976.

SOME NEW ADVANCEMENTS IN THE PHYSICAL APPLICATIONS OF ERGODIC THEORY

L.A.Bunimovich[1)]

The main concepts of ergodic theory appeared in the works of the statistical physics founders L.Boltzmann and J.Gibbs. It were the problems of statistical mechanics (first of all the problem of irreversibility) that gave an impetus to arising of this theory. Therefore advancements in these problems are indexes of maturity of the whole theory.

Later the main problem of ergodic theory was formulated much wider: you have to explain the chaotic behaviour of deterministic systems (in particular of trajectories of systems of ordinary differential equations) and you have to explain in what sense they resemble random processes.

The founders of ergodic theory have assumed that chaotic behaviour of deterministic systems is connected with very large number of their degrees of freedom. But after the work of J.Hadamard, G.Hedlund and E.Hopf on geodesic flows on surfaces of negative curvature it became clear that stochasticity can arise in systems with small number of freedom degrees owing to their local exponential instability.

Essential progress in the research of such systems was achieved in the middle of sixties, when in works by D.V.Anosov [1] and S.Smale [17] were defined the wide classes of smooth dynamical systems (Anosov systems and more general class of systems with axiom A) with a local exponential instability. In particular, geodesic flows on manifolds of negative curvature belong to these classes.

Nevertheless the main model of statistical mechanics - a gas of hard spheres with elastic interaction - does not belong to these classes. The cause is connected with discontinuity of correspondent dynamical system. The fact of local exponential instability of a hard spheres gas was stated with help of heuristical considerations by the soviet physicist N.S.Krylov [11]. Boltzmann's hypothesis on ergodicity of a hard spheres gas was proved by Ya.G.Sinaĭ [14]. Moreover in [14] was shown that a hard spheres gas is a K-system. These results are most remarkable among all rigorous results of statistical physics.

A new branch of ergodic theory appeared in last years. It can be called ergodic theory of dissipative systems. The central concept here is the notion of strange attractor introduced in 1971 in famous paper by D.Ruelle and F.Takens. The main model for applications of ergodic theory of conservative systems is the hard spheres gas. For ergodic theory of dissipative systems this role is played by the flow of viscous

1) P.P.Shirshov Institute of Oceanology , Academy of Sciences USSR , 117218 Moscow , USSR.

fluid where internal friction dissipates the kinetic energy into heat. From the very beginning attempts were made to connect the notion of strange attractor with the onset of turbulence in the flow of viscous fluid. Strange attractors have a complicated topological structure and the trajectories of the corresponding dynamical system resemble qualitatively the trajectories of fluid particles in a turbulent flow. Therefore arises a natural question: if the behaviour of dynamical system on a strange attractor is an adequate image of turbulence. For answering this question it is necessary to investigate stochastic properties of consideral dynamical system.

Ergodicity is not a crucial property for physical applications. The first important property for explaining relaxation processes is mixing. This property means that any nonequilibrium distribution, which is given by a density with respect to an invariant distribution, has to evolve towards that distribution, i.e. in a mixing system fluctuations are relaxed.

Many arised problems in physical applications demand a studying of more delicate properties of dynamical system. The most important among these properties is a speed of mixing (which is called by physicists "a speed of correlation functions decay"). One needs this fact, for example, for the proof of transport coefficients existence in nonequilibrium statistical physics. From the other side the speed of correlation function decay gives some information about a power spectrum of physical system.

The main tool for estimating of the speed of mixing is a Markov partition [15 , 4] of dynamical system's under consideration phase space. There exists a canonical method for constructing of Markov partitions for systems with axiom A [4] . So it is that a construction is much more complicated in presence of discontinuities. In all existing examples for construction of Markov partitions some subtle properties of dynamical systems are used. (For example, the method of paper [6] can not be applied directly to Lorentz gas in dimension $d > 2$).

It is the main purpose of this report to demonstrate a new general method of Markov partitions construction for hyperbolic dynamical systems with discontinuities. For the simplicity I shall discuss this construction for some important examples which are generalization of Lorentz attractors and Lorentz gas.

Only several years ago R.Balescu has written in his book [2] that for real systems there exist no theorems concerning a speed of relaxation and there exist no understanding of mechanisms of this process. I shall show that an essential step in this direction is now done.

The most simple first example is given for illustration of the idea of a construction. Let us consider a piecewisely smooth one-to-one hyperbolic mapping T of a square S. We shall denote by ∂S the boundary of S. Such linear transformation can be defined by a 2 x 2 unimodular matrix with integer elements. (For definiteness one can consider the matrix $\begin{pmatrix} 2 & 1 \\ 1 & 1 \end{pmatrix}$). Then (0,0) is the hyperbolic stationary point. In any point of S (besides the edge-points) there exist contracting and expanding local fibers (c.l.f. and e.l.f.) which are straight segments in this case. The end-points of these fibers belong to $S_- = \bigcup_{k=0}^{\infty}(T^{-k}\partial S) = \bigcup_{k=0}^{\infty} S_{-k}$ and $S_+ = \bigcup_{k=0}^{\infty}(T^{k}\partial S) = \bigcup_{k=0}^{\infty} S_k$ correspondingly.

We shall denote by $\hat{S}_1^1(i)$, $i = 1, 2, \ldots, m_{11}$, the set consisted of all such l.c.f. which intersect two arbitrary components of the set S_1. (Here m_{11} is a number of pairs of different components of the set S_1), Let $\hat{S}_1^1 = \bigcup_i \hat{S}_1^1(i)$. Consider for every l.c.f. which belongs to $\hat{S}_1^1$ all the largest connected subsets whose end-points belong to S_1. Let $\tilde{S}_1^1$ be the set which consists of all these subsets. We can consider the natural decomposition $\tilde{S}_1^1 = \bigcup_i \tilde{S}_1^1(i)$ where some of sets $\tilde{S}_1^1(i)$, $i = 1, 2, \ldots, m_{11}$, can be empty. Now we shall introduce in analogous way the sets $\hat{S}_1^2(i) \setminus \tilde{S}_1^1$, $i = 1, 2, \ldots, m_{12}$ and construct with their help the sets $\tilde{S}_1^2(i)$, $i = 1, 2, \ldots, m_{12}$. Again some of these sets can be empty. Further we shall consider $\hat{S}_2^2(i) \setminus (\tilde{S}_1^1 \cup \tilde{S}_1^2)$, $i = 1, 2, \ldots, m_{22}$ and construct the set $\tilde{S}_2^2$ and so on. In the same way we can construct the sets $\tilde{S}_{-n}^{-1}$ for all pairs of positive integers n and l.

We shall consider now the measureable partition ξ of the square S with elements $A_{n_1 n_2}^{l_1 l_2}(i,j) = \tilde{S}_{n_1}^{l_1}(i) \cap \tilde{S}_{n_2}^{l_2}(j)$; $n_1, n_2, l_1, l_2 > 0$ $i = 1, 2, \ldots, m_{n_1 l_1}$; $j = 1, 2, \ldots, m_{n_2 l_2}$. It is well known that elements of Markov partition must be parallelograms. Recall that a subset D of positive measure is called a parallelogram if for every pair of points x^I, $x^{II} \in D$ the intersection $\gamma^{(s)}(x') \cap \gamma^{(u)}(x'')$ is not empty and consists precisely of one point also belonging to D. Here $\gamma^{(s)}(x)$ $(\gamma^{(u)}(x))$ is l.c.f. (l.e.f.) which contains a point x . It is not hard to proof that all the sets $A_{n_1 n_2}^{l_1 l_2}(i,j)$ are parallelograms and ξ is Markov partition.

Lorentz attractor and Belykh attractor.

Consider the following mapping **T** of the square $A = A_1 A_2 A_3 A_4$ into itself.

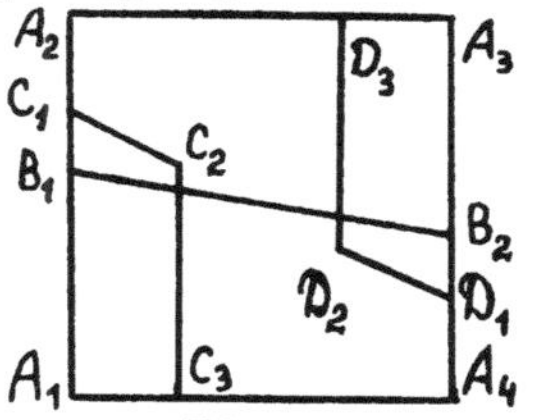

Figure 1.

The closed quadrilateral $A_1C_1C_2C_3(A_3D_1D_2D_3)$ is the image of $A_1B_1B_2A_4(A_2B_1B_2A_3)$. The points A_1 and A_3 are stationary points of hyperbolic type. Besides the segment $A_1A_4(A_3A_2)$ is contained in the stable separatrix of $A_1(A_3)$. For simplicity we shall assume that T is linear on $A_1B_1B_2A_4(A_2B_1B_2A_3)$. This mapping was considered by V.N.Belykh [3]. A limit set Λ we shall call Belykh attractor. If B_1B_2 is the midline then we obtain an attractor of Lorentz type. In order to obtain the classical Lorentz attractor we must retract the both segments C_1C_2 and D_1D_2 into two points.

Lorentz Gas.

Let us consider a random distribution of nonintersecting circles (scatterers) on the plane. On the complement of these circles' union point particles are uniformely distributed. The scatterers are fixed and every particle moves with the constant velocity and at the moments of reflections it changes its velocity according to the usual law of elastic collisions. In analogous way Lorentz gas can be defined for an arbitrary finite dimension. This model was introduced in 1905 by H.Lorentz for explaining of the electroconductivity of metals. Now it is one of the most popular models in nonequilibrium statistical physics [9].

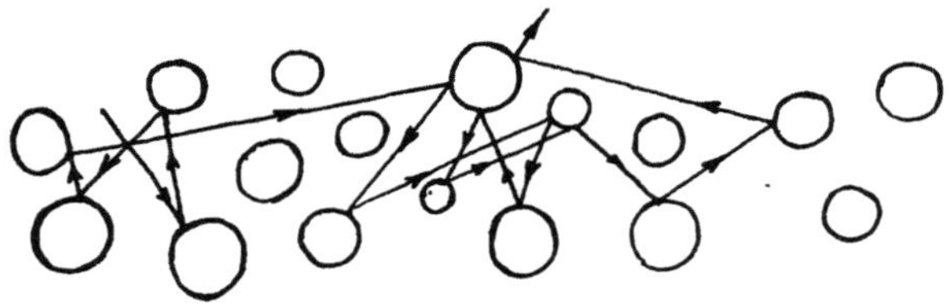

Figure 2.

We shall assume that all scatterers have one and the same radius and their configuration is periodic. Then the whole motion can be considered on a two-dimensional torus R. The phase space M of our dynamical system consists of points $x = (q, v)$, where $q = (q^1, q^2)$ are coordinates, $v = (v^1, v^2)$ is velocity of the particle. Without any loss of generality we can restrict ourselves by the case $\|v\| = ((v^1)^2 + (v^2))^{1/2} = 1$. It is easy to see that $q \in Q$, where configurational space Q is open bounded connected domain on R. Its boundary consists of a finite number of circumferences Γ_i, $i = 1, 2, \ldots, P$.

M is the unit tangent bundle over Q. Let Π be the natural projection of M onto Q. Preimage $\Pi^{-1}(q) = S^1(q)$, $q \in Q$ consists of unit vectors which are tangent to Q at $q \in Q$. The phase space is the three-dimensional open manifold with the boundary $\partial M = \bigcup_{i=1}^{p} \Pi^{-1}(\Gamma_i) = \bigcup_{i=1}^{p} \partial M_i$
We consider the framing of each Γ_i by unit normal vectors $n(q)$,

$q \in \Gamma_i$, directed inside Q. On every ∂M_i one can introduce natural coordinates (r , φ) where r is the parameter of length on every Γ_i and φ is the angle between x and n(q), q = Π (x). Let

$$M_1 = \{x \in \partial M : (x, n(q)) \geq 0, q = \Pi(x)\},$$

$$S_0 = \{x \in \partial M : (x, n(q)) = 0, q = \Pi(x)\}.$$

So M_1 is the union of P cylinders.

Liouville measure on M takes the form $d\mu = dq\, d\omega$ where dq is the usual Lebesgue measure on the plane, $d\omega$ is the Haar measure on the circumference $S^1(q) = \Pi^{-1}(q)$. One-parameter group of shifts along the trajectories is denoted by $\{S^t\}$, $-\infty < t < \infty$. It is wellknown that $\{S^t\}$ preserves the measure μ . Let $\tau(x)$, $x \in M_1$, be the nearest positive moment of reflection of the trajectory from the boundary. If $\tau(x) < C < \infty$ then one says that Lorentz gas has a finite horizon. This case was considered (for two-dimensional Lorentz gas) in papers [6, 7] . We shall also make this assumption for simplicity.

One can define the transformation of M_1 into itself by the formula $Tx = S^{\tau(x)-0}$, $x \in M_1$. It is easy to see that T preserves the measure ν where $d\nu$ = const $\cos\varphi\, dr\, d\varphi$. Here a const does not depend on Γ_i and can be chosen in such a way that ν will be normed measure. Using the language of ergodic theory one can say that the flow $\{S^t\}$ is represented as a special flow built with the help of the automorphism T of the base space M_1 and the function $\tau(x)$.

The transformation T is discontinuous on the set $S_{-1} = T^{-1}S_0$. The set S_1 consists of a finite number of smooth curves. These curves are monotonic in coordinates (r, φ) [16] . Analogously a set $S_k = T^k S_0$ consists of a finite number of smooth monotonic curves. The ends of any such curve belong to the set $\bigcup_{j=0}^{k-(\operatorname{sgn} k)1} S_j$. If k changes a sign then the corresponding curves change their monotonicity.

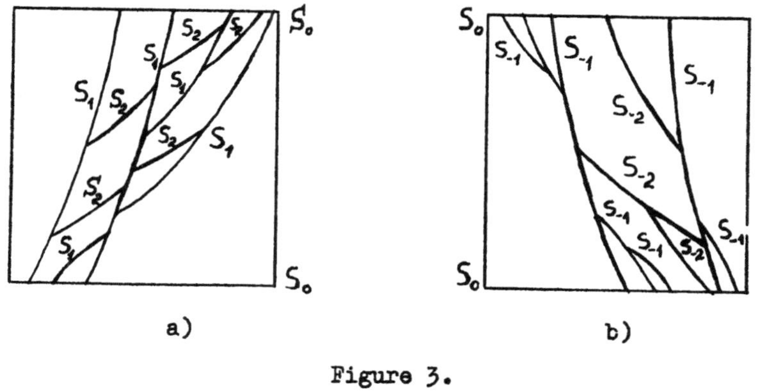

Figure 3.

So we have for Lorentz gas with finite horizon quite the same situation as for a piecewisely smooth hyperbolic automorphism of the square.

Consequently a Markov partition for Lorentz gas can be constructed in quite the same manner.

The analogous idea can be explored for construction of Markov partition for Belykh attractor. We shall describe it briefly for the case when T is linear on every component of continuity (Fig. 1). Then l.e.f. (l.c.f.) are vertical (horizontal) segments. An attractor consists of vertical segments. Let us consider for any integer k two sets $R_k = B_1B_2 \cap T^kA_2B_1$ and $Q_k = B_1B_2 \cap T^kA_4B_2$. Every point $r_i(q_i)$, $i = 1, 2, \ldots, 2^k$ of the set $R_k(Q_k)$ we shall connect by a horizontal segment $I_k^i(\mathcal{J}_k^i)$ with a nearest component of the set $T^kA_4B_2(T^kA_2B_1)$. An element $D_k^{(j)}$ $(j = 1, 2, \ldots, 2^k)$ of our Markov partition consists of all vertical segments (which belong to Λ and do not intersect B_1B_2) whose end-points belong to a pair of nearest segments I_k^i and $\mathcal{J}_k^\ell$. It is easy to see that an every set $\mathcal{D}_k^{(j)}$ is a parallelogram.

The construction of Markov partition described above can be applied to a general hyperbolic dynamical system with singularities. (Some sufficient conditions for hyperbolicity of a discontinuous dynamical systems can be found in [10]). In particular quite analogously Markov partitions can be constructed for Lorentz gas with infinite horizon in any dimension and for a hard spheres gas.

The most striking feature of our approach is that it gives (in any case for discontinuous hyperbolic systems) a Markov partition at once, i.e. without any iterative procedure. It seems that this approach is sufficiently general. It can be applied to systems with axiom A and apparently to systems with nonzero Lyapunov characteristic exponents.

As usually [4] with the help of Markov partition one can consider a symbolic dynamics. The properties of this dynamics are defined by an intrinsic nature of the consideral system. In such a way we connect some random process with our dynamical system. In order to explore various statistical properties of the dynamical system we must study regularity properties of this process. In order to do this one must explore sufficiently delicate properties of a consideral dynamical system [5, 7, 8] .

Some important results were obtained on this way recently. In particular we have shown that a speed of mixing for periodic Lorentz gas with infinite horizon has an power asymptotic. For periodic Lorentz gas with finite horizon and for Lorentz attractors a correlation function decay is asymptotically const exp $(-n^{\alpha})$, where $0 < \alpha < 1$ [7, 8] .

References

[1] Anosov,D.V.: Geodesic flows on closed Riemannian manifolds of negative curvature. Proc. Steklov Inst. of Math. 90(1967).

[2] Balescu,R.: Equilibrium and nonequilibrium statistical mechanics. Part 2. New York 1975.

[3] Belykh,V.N.: Qualitative methods of the theory of nonlinear oscillations in point systems. Gorki Univ. Press 1980.

[4] Bowen,R.: Methods of symbolic dynamics. Moscow 1979.

[5] Bunimovich,L.A. and Ya.G.Sinai: Correlation functions decay in onedimensional ecological models. Thermodynamics and kinetics of biological processes. Moscow 1980, 226-234.

[6] Bunimovich,L.A. and Ya.G.Sinai: Markov Partitions for Dispersed Billiards. Commun. Math. Phys. 72(1980), 247-280.

[7] Bunimovich,L.A. and Ya.G.Sinai: Statistical Properties of Lorentz Gas with Periodic Configuration of Scatterers. Commun. Math. Phys. 78(1981), 479-497.

[8] Bunimovich,L.A.: Statistical Properties of Lorentz Attractors. In: Bifurcations and Chaos in Dynamical Systems. London . (In press.)

[9] Hauge,E.G.: What can one learn from Lorentz models? In: Transport Phenomena. Berlin 1974.

[10] Katok,A. and I.M.Strelcyn: Invariant Manifolds for Smooth Maps with Singularities. I. Existence. Preprint Univ. of Maryland 1980.

[11] Krylov,N.S.: Works on the Foundations of Statistical Physics. Princeton Univ. Press 1979.

[12] Lorenz,E.N.: Deterministic Nonperiodic Flow. J. Atm. Sci. 20(1963), 130-141.

[13] Ruelle,D.: Thermodynamic Formalism. New York 1978.

[14] Sinaĭ,Ya.G.: On Foundation of Ergodic Hypotheses for One Dynamical System of Statistical Mechanics. Dokl. AN SSSR 153(1963), 1261-1264.

[15] Sinaĭ,Ya.G.: Markov Partitions and U-diffeomorphisms. Funct. Anal. Appl. 2(1968), 61-82.

[16] Sinaĭ,Ya.G.: Dynamical Systems with Elastic Reflections. Russian Math. Surveys 27(1970), 21-69.

[17] Smale,S.: Differentiable Dynamical Systems. Bull. Amer. Math. Soc. 73(1967), 747-817.

STATISTICAL DECISION PROCEDURES AND ERGODIC THEORY

Manfred Denker[1)]

K.Pearson [16] observed 80 years ago that for many practical statistical applications the classical approach of complete knowledge of the family of distributions determining the statistical problem is not suitable (cf. [18]). For example, testing a new medicament for good one considers a family of distributions $\{F_\theta(x) = F(x+\theta) \mid \theta \in \mathbb{R}\}$ with unknown distribution function F. This is considered to be the starting point of a new field of interest in statistics which is called nonparametric statistics. Obviously, the classical problem in statistical decision theory of finding optimal decision procedures is of no importance in this context. Other criteria for goodness of a decision procedure were developed, the most important one being the weak convergence of the test statistics to some known probability law. Most of the authors working on this last problem assumed independence of the observed samples. During the last thirty years, however, this condition has been relaxed to m-dependent processes, Markov chains and finally to weakly dependent (and stationary) processes. At this point Ergodic Theory comes in. In probabilistic terminology Ornstein's weak Bernoulli (w.B.) property is just the notion of absolute regularity, known for quite a while in probability (cf. [11]). Thus, for these w.B. processes the results from statistics on absolutely regular processes are applicable. However, having a closer look at these questions we shall see in a while that from the viewpoint of dynamical systems the statistical results obtained so far (except in [6]) are not precisely what is needed. I shall explain this by an example.

Transformations on the unit interval, sometimes called first order recurrence, serve as a model for various purposes. In population dynamics of species with nonoverlapping generations the transformations

$$T(x) = x \exp a(1-\frac{x}{b}) \quad \text{or} \quad T(x) = ax(1-x)$$

are well known in biology (cf. [1], [12], [14], [17]). In economics a (very simple) delayed price evolution model leads to the implicite equation $g(Tx) = f(x)$ (cf. [8]). Similar examples and others from physics and medicine can be found in the literature. In all these examples the transformation itself (the size of the following generation or the prize after one time unit) is measured and thus one obtains a sequence $x, T(x), T^2(x), \ldots, T^{n-1}(x)$ of n observed iterates of the

1) Partially supported by the Deutsche Forschungsgemeinschaft.

transformation. In order to see how it is linked to statistical methods let us assume some properties of T and then formulate the situation in probabilistic language.

Suppose the transformation $T : [0,1] \rightarrow [0,1]$ satisfies the conditions of Lasota and Yorke ([13]):

(1) there exists a partition $0 = a_o < a_1 < \ldots < a_p = 1$ such that $T|\,]a_{i-1},a_i[$ is monotone and has a C^1-extension to $[a_{i-1},a_i]$ $(i=1,\ldots,p)$.

(2) $1 < \inf|T'|$

(3) T' is Lippschitz continuous.

(More generally one can think of a transformation with the Hofbauer-Keller condition ([9]))- Let us assume moreover that

(4) T is weakly mixing with respect to the invariant probability P.

Denote by α the natural partition of T given by the points a_i $(0 \leq i \leq p)$. Bowen [4] showed that the natural extension of (T,α) is w.B. which in probabilistic language means that

$$E \sup_{B\in(\alpha)_n^\infty} |P(B|(\alpha)_{-\infty}^{o}) - P(B)| = \beta(n) \downarrow 0$$

(this is just the absolute regularity condition). It is straightforward to see that $\beta(n)$ decreases to 0 exponentially fast. Yoshihara [19] has considered nonparametric statistics of absolutely regular processes; his results can, however, only be applied if the process (T,α) is observable, which means that the natural partition α has to be known. Since one can only hope to know an orbit segment $x, Tx,\ldots,T^{n-1}x$, we have to look for its probabilistic interpretation. It is immediately to be seen that the variation of T on an element of $(\alpha)_{-n}^{o}$ decreases exponentially implying that

$$E|T - E(T|(\alpha)_{-n}^{o})|^q$$

also decreases exponentially fast (for every $q \in \mathbb{N}$). This property shows that the orbit section is a functional of the absolute regular process (T,α) in the sense of Ibragimov-Linnik [10], or that T is a sequential function of (T,α) as defined in [5].

Consider now the problem of detecting a difference in the invariant distributions of two independent transformations S and T (fulfilling conditions (1)-(4)) when knowing orbit segments $x, T(x),\ldots,T^{n-1}(x)$ and $y,S(y),\ldots,S^{m-1}(y)$. It has been shown in [6] that the following procedures are well suitable, known as the Wilcoxon 2-sample and the

Lehmann test.

(a) The Wilcoxon test W is good when detecting a difference in location (see the beginning). Let $h(x,y) = 1_{\{x>y\}}$ and set

$$W = \sum_{j=0}^{n-1} \sum_{k=0}^{m-1} h(T^j x, S^k y) + \frac{n(n+1)}{2}$$

Obviously, if the distributions are equal, then $EW = \frac{n}{2}(n+m+1)$.

(b) The Lehmann test is mainly used to compare two density functions. Let F (resp. G) denote the distribution function of T (resp. S) and let

$$\Delta(F,G) = \int (F(z) - G(z))^2 \, d\,\frac{F+G}{2}(z)$$

If $F = G$, then $\Delta(F,G) = 0$. The statistic is given by

$$L = \frac{1}{2}\binom{n}{2}^{-1}\binom{m}{2}^{-1} \sum_{\substack{0\le i<j\le n-1\\ 0\le k<l\le m-1}} \left(h(T^i x, T^j x; S^k y, S^l y) - \frac{1}{3}\right)$$

where

$$h(x_1,x_2,y_1,y_2) = \begin{cases} 1 & \text{if } x_1 \vee x_2 < y_1 \wedge y_2 \text{ or } y_1 \vee y_2 < x_1 \wedge x_2 \\ 0 & \text{otherwise.} \end{cases}$$

It is not difficult to see that

$$\iiiint h(x_1,x_2,y_1,y_2)\,dF(x_1)\,dF(x_2)\,dG(y_1)\,dG(y_2) = \frac{1}{3} + 2\Delta(F,G),$$

hence also in our case $\lim_{n,m\to\infty} EL = 0$ if $F = G$.

In order to have good tests one has to prove that after a suitable normalization the statistics approach a weak limit when $n,m \to \infty$. In [6] we proved a theorem which gives a normal law as a limit and which will be explained now:

(5) Let $(Z_n^i)_{n\in\mathbb{Z}}$ $(i=1,2)$ be two absolutely regular and stationary processes, (Z_n^1) independent of (Z_n^2), with coefficients

$$\beta(u) = O(u^{-\frac{\delta}{1-\delta}}) \text{ for some } 0 < \delta < 1,$$

having a compact state space E.

(6) Let $g,f : E^{\mathbb{N}} \to \mathbb{R}$ be Hölder-continuous with exponent s and some constant C. We shall consider the stationary processes

$$X_n := f(Z_n^1, Z_{n+1}^1, \dots)$$

$$Y_m := g(Z_m^2, Z_{m+1}^2, \dots)$$

Let $g^1, f^1 : E^1 \to \mathbb{R}$ be given by conditional expectations

$$g^l(u_1 \ldots u_l) = E(g \mid Z_j^2 = u_j \quad 1 \le j \le l)$$

$$f^l(u_1 \ldots u_l) = E(f \mid Z_j^1 = u_j \quad 1 \le j \le l).$$

It follows now that for a.e. $u \in E^N$

$$|g(u) - g^l(u^l)| \le C\, 2^{-ls} \quad (u^l = (u_1 \ldots u_l))$$

and similarly for f.

Therefore

$$X_n^l = f^l(Z_n^1, Z_{n+1}^1, \ldots, Z_{n+l-1}^1) \quad \text{and}$$

$$Y_m^l := g^l(Z_m^2, Z_{m+1}^2, \ldots, Z_{m+l-1}^2)$$

are "good" approximations of the processes (X_n) and (Y_m) and absolutely regular themselves.

Let h be a function defined on $\mathbb{R}^q$. The ε-oscillation of h in $u \in \mathbb{R}^q$ is defined by

$$\operatorname{osc}(h,\varepsilon,u) = \sup\{|h(v)-h(w)| \mid \|v-u\|, \|w-u\| < \varepsilon\}.$$

This is a measurable (semicontinuous) function, so the mean oscillation of h with respect to a product measure Q can be defined as

$$\operatorname{osc}(h,\varepsilon) := \int \operatorname{osc}(h,\varepsilon,u)\, Q(du) \quad \text{(cf. [15])}.$$

Here we assume that

(7) h has bounded Q-variation in the sense that

$$\sup_{\varepsilon>0} \frac{\operatorname{osc}(h,\varepsilon)}{\varepsilon^r} =: M < \infty \quad \text{for some } r > 0$$

where Q denotes the product measure of m_1 copies of the distribution F of X_1 and m_2 copies of G the distribution of Y_1 (note that $q = m_1 + m_2$).

The condition on h is fulfilled if F and G are absolutely continuous and $F(b) - F(a) \le C_1|b-a|^r$ for some $C_1 > 0$ (similarly for G) and if h assumes only finitely many values and its set of discontinuities is contained in a finite union of hyperplanes in $\mathbb{R}^q$.

Lemma 1: [6]

Conditions (6) and (7) imply that for any $t_1,\ldots,t_{m_1},s_1,\ldots,s_{m_2}$

$$E|h(X_{t_1},\ldots,X_{t_{m_1}},Y_{s_1},\ldots,Y_{s_{m_2}}) - h(X_{t_1}^l,\ldots,X_{t_{m_1}}^l,Y_{s_1}^l,\ldots,Y_{s_{m_2}}^l)|$$

$$\le M\, C^r\, 2^{-lrs} + 2\|h\|_\infty\, \beta(\min\{|t_i-t_j|, |s_\nu-s_\mu| \mid i\neq j, \mu\neq\nu\} - l) \cdot (q-1)$$

<u>Theorem 2:</u> [6]

Let conditions (5)-(7) be satisfied. Assume that h is symmetric and $\int\ldots\int h(x_1\ldots x_{m_1}y_1\ldots y_{m_2}) \prod_{i=1}^{m_1} dF(x_i) \prod_{j=1}^{m_2} dG(y_j) = 0$.
Define

$$h_1(u) = \int\ldots\int h(u,u_2,\ldots,u_{m_1},v_1,\ldots,v_{m_2})dF(u_2)..dF(u_{m_1})dG(v_1)..dG(v_{m_2})$$

and

$$h_2(v) = \int\ldots\int h(u_1,\ldots,u_m,v,v_2,\ldots,v_{m_2})dF(u_1)..dF(u_{m_1})dG(v_2)..dG(v_{m_2}).$$

Then

$$\sigma_1^2 = E(h_1(X_1))^2 + 2 \sum_{t\geq 2} E\, h_1(X_1)h_1(X_t) < \infty$$

and

$$\sigma_2^2 = E(h_2(Y_1))^2 + 2 \sum_{t\geq 2} E\, h_2(Y_1)h_2(Y_t) < \infty .$$

If $0 < \alpha = \lim \frac{n}{n+m} < 1$ and $\sigma^2 = \alpha^{-1}m_1^2\sigma_1^2 + (1-\alpha)^{-1}\, m_2^2\sigma_2^2 > 0$,
then

$$\frac{\sqrt{n+m}}{\sigma} \binom{n}{m_1}^{-1} \binom{m}{m_2}^{-1} \sum_{\substack{1\leq t_1<..<t_{m_1}\leq n \\ 1\leq s_1<..<s_{m_2}\leq m}} h(X_{t_1},\ldots,Y_{s_{m_2}}) \to N(0,1)$$

weakly.

The function $h = 1_{\{x-y>o\}}$ clearly satisfies (7), if F and G are absolutely continuous and satisfy

$$\max\{F(b)-F(a),\ G(b)-G(a)\} \leq C|b-a|^r.$$

It follows that for the Wilcoxon 2-sample statistic in the case of transformations of the Lasota-Yorke type we are discussing, the Theorem 2 holds true. For the Lehmann statistic it is not so obvious. It may happen (and it certainly is true when $F = G$), that $\sigma^2 = 0$. In this case the limiting distribution still is an open question, however, in the remaining part of this paper I shall prove a similar theorem which is the first step to solve this problem.

Before doing this let us see whether the results obtained so far really can be applied. We (G.Keller and myself) did some computer simulations of which I shall discuss three examples. For applications it is necessary to estimate the variance σ^2 under the hypothesis $F = G$. This is done by a similar method as explained before. The details can be found in [6], too.

a) Take $T(x) = 1{,}634 \cdot x \bmod 1$ and $S = T$. The transformations were iterated 350 times with starting values $x = 0{,}802577$ and $y = 0{,}369254$, thus obtaining $x, T(x), T^2(x),\ldots,T^{350}(x)$ and $y, S(y), S^2(y),\ldots$ $\ldots, S^{350}(y)$. The normalized values of the Lehmann- and Wilcoxon statistic are 0,999577 and 1,018814 (the ideal value would be 1). For an error probability of $\alpha = 0{,}01$ under the hypothesis of $F = G$ the Wilcoxon 2-sample test would accept $F = G$ if the Wilcoxon value (in our case 1,018814) lies in the interval [0,96;1,04]. Here one has to use the estimation of the asymptotic variance σ^2. For the Lehmann statistic these estimates are very bad (because they are almost zero) due to the theoretical value. This means that we cannot control the error probability, although the Lehmann value seems to be very precise (even in the case where the asymptotic law would be normal).

b) Let $T(x) = 9,367\, x \bmod 1$ and $S(x) = 1,401\, x \bmod 1$ and consider again 350 iterates with starting values $x = 0,687332$ for T and $y = 0,065839$ for S. Under the hypothesis the Wilcoxon test accepts in the interval [0,96;1,04] assuming an error of $\alpha = 0,01$; the Wilcoxon value is 0,862262, thus rejecting the hypothesis $F = G$. The Lehmann value in this case is 1,022138. (Here the asymptotic variance estimation is rather big and as a first critical region in which one should not trust too much, becomes [0,9861;1,0139]. This indicates at least that after developing the theory one could expect good results in applications.)

c) In our third example let

$$T(x) = 0,983\ (1-2|x-\tfrac{1}{2}|)$$

and

$$S(x) = 1 - 4(x-\tfrac{1}{2})^2 .$$

Again with 350 iterates and starting values $x = 0,685324$ and $y = 0,882563$ the Wilcoxon value is 0,999201 and the Lehmann value 1,006861. With an error probability of $\alpha = 0,01$ we obtain in the Wilcoxon case the critical region [0,953;1,047]. The Wilcoxon test accepts although the distributions are different! However, this is not surprising since the distributions differ not very much in location, and it is well known from the case of independent samples that the Wilcoxon test is bad in such situations. With the same caution as discussed in the last example we can get an idea of the critical region for the Lehmann test. We obtain here [0,999;1,001], thus the Lehmann test would reject. This example shows that considering the degenerate case $\sigma^2 = 0$ of Theorem 2 is very important.

The best known statistics in the degenerate case are tests of fit, especially the χ^2-test, first considered by Pearson [16]. They are also of particular interest for interval transformations fulfilling conditions (1)-(4). Let T be such a transformation, and consider the problem of detecting a difference in its invariant distribution compared to some fixed probability given by $\varphi(x)dx$ ($\varphi \geq 0$, $\int \varphi(x)dx = 1$). Here we can use a χ^2-test of fit. Divide [0,1] into disjoint intervals I_k $(1 \leq k \leq K)$ and set

$$p_k := \int_{I_k} \varphi(x)dx.$$

Then define

$$h(x,y) = \sum_{k=1}^{K} p_k^{-1}(1_{I_k}(x)-p_k)(1_{I_k}(y)-p_k)$$

and form the χ^2-test

$$\chi_N^2 = N^{-1} \sum_{0\le t<s\le N-1} h(T^t(x),T^s(x)).$$

Under the hypothesis $F(x) = \int^x \varphi(u)du$ (F the distribution of the invariant measure for T), we have $\int h(x,y)\,\varphi(x)dx = 0$ for all y, i.e., $h_1(u) = 0$ for all u. This means that χ_N^2 rejects for large values and accepts for small values. It also shows that the statistic is degenerate, since σ^2 of Theorem 2 equals zero. (Theorem 2 can also be applied to the one-sample case.) Hence we are looking for other normalizing procedures hoping for a good limiting procedure.

For the remaining part of this paper let (Z_n) be an absolutely regular process satisfying condition (5) with $\beta(n) = O(n^{-8})$. Also let $X_n = f(Z_n,Z_{n+1},\ldots)$ be a stationary functional of (Z_n) for which (6) holds, and we shall consider a symmetric function $h : \mathbb{R}^2 \to \mathbb{R}$ of bounded P-variation in the sense of (7). In addition, we shall assume that (8): $F(b) - F(a) \le C_1|b-a|^r$ $(a < b)$ and that the distributions of (X_o,X_n) and (X_n,X_o) $(n \ge 1)$ have measure zero on the diagonal. F denotes the distribution of X_o and, in particular, F is continuous and F^{-1} is uniquely determined.

<u>Lemma 3</u>:

The process $Y_n = F \circ X_n$ and the function $\hat{h}(x,y) = \begin{cases} h(F^{-1}x,F^{-1}y), & x\neq y \\ 0, & x=y \end{cases}$ have the following properties:

(a) The distribution of Y_o is Lebesgue measure λ on $[0,1]$.

(b) The distributions of the processes $(h(X_n,X_m))_{n\neq m}$ and $(\hat{h}(Y_n,Y_m))_{n\neq m}$ are the same.

(c) $|Y_o - E(Y_o|\sigma(Z_l\,|\,|l|\le n))| \le c^r C_1 2^{-srn}$

<u>Proof</u>:

(a) $P(Y_o<t) = P(X_o < F^{-1}t) = F(F^{-1}t) = t$

(b) Since F is continuous we may assume that h vanishes on the diagonal. So the equality follows from $\hat{h}(Y_n,Y_m) = h(F^{-1}FX_n, F^{-1}FX_m) = h(X_n X_m)$.

(c) Because f is Hölder-continuous and because of (8) we have

$$|F \circ f(Z_o,Z_1,\dots) - Ff^l(Z_o,Z_1,\dots,Z_{l-1})|$$
$$\le C_1|f(Z_o,Z_1,\dots) - f^l(Z_o,Z_1,\dots,Z_{l-1})|^r$$
$$\le C_1 c^r\, 2^{-rsl}$$

and this remains true for the conditional expectation.

Lemma 3 is quite important since it shows that we can restrict the problem of the limiting distribution to processes with uniformly distributed one-dimensional marginals. In this case and under the general assumptions made we can use a theorem of Berkes and Philipp [2]:

Define

$$g_n(t) = 1_{\{Y_n<t\}} - t \qquad 0\le t\le 1$$

and

$$\Gamma(s,t) = Eg_o(t)g_o(s) + \sum_{n\ge 1} E(g_o(t)g_n(s) + g_o(s)g_n(t)) \qquad (0\le s,t\le 1)$$

It is well known that $\Gamma(s,t)$ is convergent because of Lemma 3(c) and it is immediately to be seen that Γ is a positive definite function since the process is stationary:

$$\sum_{k,j=1}^{q} \alpha_k\Gamma(t_k,t_j)\alpha_j = \lim_{N\to\infty} \frac{1}{N} E\Big(\sum_{k=1}^{q}\sum_{n=o}^{N-1} g_n(t_k)\alpha_k\Big)^2 \ge 0.$$

Since Γ is positive definite there exists a Gaussian random variable $B(t)$ on $C([0,1])$ with $E\,B(t) = 0$ and covariance operator

$$E\,B(t)\,B(s) = \Gamma(t,s) \qquad 0\le t,s\le 1,$$

especially $E\,B(0)^2 = E\,B(1)^2 = 0$, hence $B(0) = B(1) = 0$. The result of Berkes and Philipp says that the empirical process of (Y_n) obeys a central limit theorem, i.e., for

$$\tilde{Y}_n(t) = \frac{1}{\sqrt{n}} \sum_{k=o}^{n-1} (1_{\{Y_k<t\}} - t)$$

we have that the distributions of $\tilde{Y}_n$ in $D([0,1])$ converge weakly to B. Here $D([0,1])$ denotes the function space of all functions on $[0,1]$ with no discontinuities of the second kind (left continuous and right limits) equipped with the Skorohod topology (cf. [3]).

Next we have to define a multiple stochastic integral with respect to B. Let $\|h\|_\infty$ denote the supremum over all L_∞-norms of h with respect to the distributions of (X_o,X_n) and (X_n,X_o) for $n \ge 0$. Define

$H = \{h : \mathbb{R}^2 \to \mathbb{R} \mid \|h\|_\infty < \infty\}$ and similarly $\hat{H}$ for the process (Y_n). Then both spaces are Banach spaces.

For a step function $\hat{h} \in \hat{H}$ with values $h_{c_i c_j}$ on the rectangle $]c_{i-1}, c_i] \times]c_{j-1}, c_j]$ where $0 = c_o < c_1 < \ldots < c_q = 1$ define

$$Z(\hat{h}) := \sum_{i,j=1}^{q} h_{c_i c_j} (B(c_i) - B(c_{i-1}))(B(c_j) - B(c_{j-1})).$$

This is a well-defined stochastic integral which has to be extended to $\hat{H}$.

Lemma 4:

There exists a constant C_2 such that for any step function $\hat{h} \in \hat{H}$ we have

$$E\, Z(\hat{h})^2 \le C_2 \|\hat{h}\|_\infty^2,$$

Proof:

Let $\hat{h}$ have the finite number of values $h_{c_i c_j}$ on $]c_{i-1}, c_i] \times]c_{j-1}, c_j]$ and set

$$\gamma_{c_i c_j} = \Gamma(c_i, c_j) - \Gamma(c_i, c_{j-1}) - \Gamma(c_{i-1}, c_j) + \Gamma(c_{i-1}, c_{j-1})$$

$$= \sum_{n=o}^{\infty} (P(c_{i-1} < Y_o \le c_i,\ c_{j-1} < Y_n \le c_j) -$$

$$- (c_i - c_{i-1})(c_j - c_{j-1})) +$$

$$+ \sum_{n=1}^{\infty} (P(c_{i-1} < Y_n \le c_i;\ c_{j-1} < Y_o \le c_j) - (c_i - c_{i-1})(c_j - c_{j-1}))$$

Then a simple calculation shows that

$$E(\iint \hat{h}(x,y)\, dB(x) dB(y))^2 =$$

$$= \sum h_{c_{i_1} c_{i_2}} h_{c_{j_1} c_{j_2}} E((B(c_{i_1}) - B(c_{i_1 - 1}))(B(c_{i_2}) - B(c_{i_2 - 1}))(B(c_{j_1}) -$$

$$- B(c_{j_1 - 1}))(B(c_{j_2}) - B(c_{j_2 - 1})))$$

$$= \sum h_{c_{i_1} c_{i_2}} h_{c_{j_1} c_{j_2}} (\gamma_{c_{i_1} c_{i_2}} \gamma_{c_{j_1} c_{j_2}} + \gamma_{c_{i_1} c_{j_1}} \gamma_{c_{i_2} c_{j_2}} +$$

$$+ \gamma_{c_{i_1} c_{j_2}} \gamma_{c_{i_2} c_{j_1}})$$

$$= 3(\int \hat{h}(x,x) dx + \iint \hat{h}(x,y) dx dy)^2 + (\sum_{n \ge 1} \int \hat{h}\, dP_{(Y_o, Y_n)} - \int \hat{h}\, d\lambda^2$$

$$+ \sum_{n \ge 1} \int \hat{h}\, dP_{(Y_n, Y_o)} - \int \hat{h}\, d\lambda^2)^2 +$$

$$+ \sum_{n,m\geq 1} \iiiint \hat{h}(x,u)\hat{h}(y,v)d(P_{(Y_o,Y_n)} - \lambda^2)(x,y)d(P_{(Y_o,Y_m)}-\lambda^2)(u,v)$$

$$+ \sum_{n,m\geq 1} \iiiint \hat{h}(x,u)\hat{h}(y,v)d(P_{(Y_o,Y_n)}-\lambda^2)(x,y)d(P_{(Y_m,Y_o)}-\lambda^2)(u,v)$$

$$+ \sum_{n,m\geq 1} \iiiint \hat{h}(x,u)\hat{h}(y,v)d(P_{(Y_n,Y_o)}-\lambda^2)(x,y)d(P_{(Y_o,Y_m)}-\lambda^2)(u,v)$$

$$+ \sum_{n,m\geq 1} \iiiint \hat{h}(x,u)\hat{h}(y,v)d(P_{(Y_n,Y_o)}-\lambda^2)(x,y)d(P_{(Y_m,Y_o)}-\lambda^2)(u,v)$$

$$+ \sum_{n,m\geq 1} \iiiint \hat{h}(x,u)\hat{h}(v,y)d(P_{(Y_o,Y_n)}-\lambda^2)(x,y)d(P_{(Y_o,Y_m)}-\lambda^2)(u,v)$$

$$+ \sum_{n,m\geq 1} \iiiint \hat{h}(x,u)\hat{h}(v,y)d(P_{(Y_o,Y_n)}-\lambda^2)(x,y)d(P_{(Y_m,Y_o)}-\lambda^2)(u,v)$$

$$+ \sum_{n,m\geq 1} \iiiint \hat{h}(x,u)\hat{h}(v,y)d(P_{(Y_n,Y_o)}-\lambda^2)(x,y)d(P_{(Y_o,Y_m)}-\lambda^2)(u,v)$$

$$+ \sum_{n,m\geq 1} \iiiint \hat{h}(x,u)\hat{h}(v,y)d(P_{(Y_n,Y_o)}-\lambda^2)(x,y)d(P_{(Y_m,Y_o)}-\lambda^2)(u,v)$$

It follows from Proposition 13 in [6], Lemma 1 and Lemma 3 that

$$(\int \hat{h}\ d\ \lambda^2)^2 \leq \|\hat{h}\|_\infty^2 \quad \text{and} \quad |\int \hat{h}(x,x)dx| \leq \|\hat{h}\|_\infty^2 .$$

Using Lemma 1 and Lemma 1 in [19] it follows that

$$|\int \hat{h}\ dP_{(Y_o,Y_n)} - \int \hat{h}\ d\ \lambda^2| = O(\|\hat{h}\|_\infty\ \beta([\tfrac{n}{3}]))$$

and

$$|\int \hat{h}\ dP_{(Y_n,Y_o)} - \int \hat{h}\ d\ \lambda^2| = O(\|\hat{h}\|_\infty\ \beta([\tfrac{n}{3}])),$$

hence the second sum is of order $\|\hat{h}\|_\infty^2$ since $\sum_n \beta([\frac{n}{3}]) < \infty$.

Again, applying Lemma 1 and Proposition 12 in [6] twice, we see that

$$|\iiiint \hat{h}(x,u)\hat{h}(y,v)d(P_{(Y_o,Y_n)}-\lambda^2)(x,y)d(P_{(Y_o,Y_m)}-\lambda^2)(u,v)|$$

$$= O(\|\hat{h}\|_\infty^2\ \beta([\tfrac{n}{3}])\beta([\tfrac{m}{3}]))$$

and so the third sum is also of order $\|\hat{h}\|_\infty^2$. Similarly, the other terms are estimated to prove the lemma.

Lemma 4 shows that a stochastic integral

$$Z(\hat{h}) = \iint \hat{h}(x,y)dB(x)dB(y) \qquad (\hat{h} \in \hat{H})$$

is well defined by approximation with step functions, and satisfies

$$E\ Z(\hat{h})^2 \leq C_2\|\hat{h}\|_\infty^2.$$

It should be remarked here that it is also possible to define a stochastic integral for functions $\hat{h}$ with finite norm

$$\|\hat{h}\|_{2+\delta} = \sup_{n\geq 0} \max\left((E|\hat{h}(Y_0,Y_n)|^{2+\delta})^{\frac{1}{2+\delta}}, (E|\hat{h}(Y_n,Y_0)|^{2+\delta})^{\frac{1}{2+\delta}} \right)$$

in a similar way.

Now, let $\hat{h}: \mathbb{R}^2 \to \mathbb{R}$ be a symmetric, square integrable function (each symmetric $\hat{h} \in \hat{H}$ is such a function!). The operator A defined by

$$A\, g(y) = \int \hat{h}(x,y)g(x)dx \qquad (g \in L_2(\lambda))$$

is Hilbert-Schmidt. Denote by $\{\lambda_k\}$ the sequence of eigenvalues (which satisfy $\sum \lambda_k^2 < \infty$) and by $\{g_k\}$ corresponding orthogonal normed eigenvectors. Then one has a representation

$$\hat{h}(x,y) = \sum \lambda_k g_k(x) g_k(y) \quad \text{in} \quad L_2(\lambda^2).$$

Assuming that $\|\hat{h}\|_\infty$, $\|g_k\|_\infty < \infty$ for all k, we can proceed similarly to Lemma 8 in [7]: Using the estimates of Lemma 4 and the hypothesis on the joint distributions in condition (8) one shows easily that for $\hat{h} \in \hat{H}$ with $\hat{h}(x,x) \equiv 0$ $(x \in \mathbb{R})$ the random variables $\iint \hat{h}(x,y)dB(x)dB(y)$ and $\sum \lambda_k[(\int g_k dB)^2-1]$ have the same distribution.

We have almost proved

<u>Theorem 5:</u>

Under the assumptions (5)-(8) the random variables

$$(*) \qquad \frac{2}{N} \sum_{0\leq n<m\leq N-1} h(X_n,X_m) - h_1(X_n) - h_2(X_m) + \iint h(x,y)dF(x)dF(y)$$

converge weakly to the variable

$$\sum_{k=1}^{\infty} \lambda_k[(\int g_k(x)dB(x))^2-1],$$

if the stochastic integrals are well defined.

<u>Proof:</u>

If h is a step function the theorem follows from the result of Berkes and Philipp [2] and the remarks above. For arbitrary $\hat{h} \in \hat{H}$ we can approximate by step functions, since it follows from Lemma 1 and Proposition 12 in [6] that the second moment of $(*)$ is of order $\|h\|_\infty^2$. Then, using the definition of the stochastic integral, one obtains the weak convergence of $(*)$ to

$$\iint \hat{h}(x,y)\, dB(x)dB(y)$$

which implies the result by the remark before.

If in Theorem 5 (*) the summation extends over all n and m, we have a limiting distribution

$$\sum \lambda_k (\int g_k dB)^2 \text{ if } h(x,x) = \sum \lambda_j g_j(x) g_j(x) \quad \lambda\text{-a.s.}$$

Here we also have to assume that $\int h^2(x,x)dF(x) < \infty$ which for Theorem 5 did not have to be supposed. Note that $\int g_k dB$ is a Gaussian random variable with $E \int g_k dB = 0$ and

$$E(\int g_k dB)^2 = 1 + \sum_{n=1}^{\infty} \int g_k(x) g_k(y) d(P_{(Y_o Y_n)} - \lambda^2)(x,y)$$
$$+ \sum_{n=1}^{\infty} \int g_k(x) g_k(y) d(P_{(Y_n, Y_o)} - \lambda^2)(x,y).$$

In the independent case $\Gamma(s,t) = s(1-t)$ $(s \le t)$ and $P_{(Y_o Y_n)} = P_{(Y_n Y_o)} = \lambda^2$, so that $\int g_k dB$ is $N(0,1)$-distributed.
Also for $k \neq j$

$$\operatorname{Cov} \int g_k dB \int g_j dB = \sum_{n=1}^{\infty} \int g_k(x) g_j(y) d(P_{(Y_o Y_o)} - \lambda^2)(x,y).$$

This implies that the family $(\int g_k dB)$ is independent when the samples (Y_n) are independent.
Let us return to the χ^2-test with kernel

$$h(x,y) = \sum_{k=1}^{K} (1_{I_k}(x) - p_k)(1_{I_k}(y) - p_k) p_k^{-1}.$$

The operator A is given by the matrix $(a_{ij})_{1 \le i,j \le K}$ where $a_{jj} = 1-p_j$ $(1 \le j \le K)$ and $a_{ij} = -p_i^{1/2} p_j^{1/2}$ $(i \neq j)$. It is easy to see that there are K-1 non-zero-eigenvalues equal to 1 and K-1 orthonormal eigenvectors are calculated easily. In the independent sample case we obtain for the distribution of $\sum \lambda_k (\int g_k dB)^2$ just the well-known χ^2-distribution of degree K-1. For the functionals of w.B. processes we are considering in this paper that the situation is not so easy. It is clear, however, that the limiting distribution can be calculated (under the hypothesis) since one could use the transformation associated with $\varphi(x)dx$ or one could estimate the variances and covariances needed from the samples $Y_o, Y_1, \ldots$ just like in [6].

With some more effort, as indicated in [7], one can prove the analogue of Theorem 5 for the two-sample-case, and presumably also in the multi-sample case. Stronger results can be obtained for the limiting distributions when considering invariance principles. For Theorem 1 the result can be easily deduced from [6] and for Theorem 2 it is also not too difficult. The two-sample-case seems to be not so easy while the multi-sample-case causes apparently deeper problems because a (certain) multidimensional maximal inequality has to be obtained first.

<u>Literature</u>:

[1] BEDDINGTON,J.R./C.A.FREE/J.H.LAWTON: Dynamic complexity in predator-prey models framed in difference equations. Nature <u>255</u> (1975), 58-60.

[2] BERKES,I./W.PHILIPP: Almost sure invariance principle for the empirical distribution functions of mixing random variables. Z.Wahrscheinlichkeitstheorie verw.Geb. <u>41</u> (1977), 115-137.

[3] BILLINGSLEY,P.: Convergence of probability measures. John Wiley & Sons, 1968.

[4] BOWEN,R.: Bernoulli maps of an interval. Israel J.Math. <u>28</u> (1978), 298-314.

[5] DENKER,M./M.KEANE: Almost topological dynamical systems. Israel J. Math. <u>34</u> (1979), 139-160.

[6] ---/G.KELLER: On U-statistics and v.Mises' functionals for weakly dependent process, preprint.

[7] ---/G.KELLER: A note on invariance principles for v.Mises' statistics,preprint.

[8] GANDOLFO,G.: Mathematical methods and models in economic dynamics. North Holland Publ.Co., Amsterdam 1971.

[9] HOFBAUER,F./G.KELLER: Ergodic properties of invariant measures for piecewise monotonic transformation;to appear in Math.Zeitschrift.

[10] IBRAGIMOV,I.A./Y.V.LINNIK: Independent and stationary sequences of random variables. Wolters-Noordhoff Publ., Groningen 1971

[11] ---/Y.A.ROZANOV: Gaussian random processes. Springer Verl. 1978.

[12] LASOTA,A.: Ergodic problems in biology. Astérisque <u>50</u> (1977),239-250.

[13] ---/Y.YORKE: On the existence of invariant measures for piecewise monotonic transformations. Trans.AMS <u>186</u> (1973), 481-488.

[14] MAY,R.: Simple mathematical models with very complicated dynamics. Nature <u>261</u> (1976), 459-467.

[15] OXTOBY,J.C.: Measure and category. Vol.3. Springer "Graduate Texts in Mathematics".

[16] PEARSON,K.: On the criterion that a given system of deviations from the probable in the case of correlated systems of variables is such that it can be reasonable supposed to have arisen from random sampling. Phil.Mag. 5 ser. Vol.50 (1900), 157-175.

[17] RICKER,W.E.: J.Fish.Res. Bd. Can.11 (1954), 569-623.

[18] SCHEFFÉ,H.: Statistical inference in the non-parametric case. Annals of Math.Stat. <u>14</u> (1943), 305-332.

[19] YOSHIHARA,R.: Limiting behaviour of U-statistics for stationary absolutely regular processes. Z.Wahrscheinlichkeitstheorie verw. Geb. <u>35</u> (1976), 237-252.

THE SPECTRUM OF SCHRÖDINGER OPERATORS WITH ERGODIC POTENTIAL

Harald Englisch [1)]

Because of its physical significance $H = -\Delta + \tilde{\lambda} V$ acting on $L^2(R^n)$ is one of the best investigated operators in analysis. This operator describes in quantum mechanics the energy of a particle in an external potential V where $-\Delta$ gives the kinetic energy, and $\tilde{\lambda}$ is the coupling constant.

Qualitative spectral analysis classifies the measures $\langle \varphi, dE_\lambda \varphi \rangle$, where φ runs through the Hilbert space $L^2(R^n)$ (in physics the case $n = 3$ is relevant), and $H = \int \lambda \, dE_\lambda$ is the spectral decomposition.

Results of the following type are well-known in spectral analysis (for exact theorems cf. [2]):

1) $\lim_{|x| \to \infty} V(x) = \infty$

   Then the spectrum $\sigma(H)$ contains only the discrete spectrum σ_{disc}, i.e. isolated eigenvalues with finite multiplicity. The harmonic oscillator and the square wall potential with infinitely high walls are physical examples for such potentials. The last example is related to the operator $H = -\Delta$ on a compact manifold.

2) $\lim_{|x| \to \infty} V(x) = 0$

   Then $\sigma(H)$ consists of the discrete spectrum in $(-\infty, 0]$ and of the absolutely continuous spectrum $\sigma_{ac} = [0, \infty)$. The potentials formed by ionized atoms or molecules are of such type.

Now the progress in spectral analysis comes mainly from investigations of ergodic potentials V_ω, where V is a map from a measure space (Ω, B, μ) into $L^p_{loc}(R^n)$ (with sufficiently large p), which is ergodic with respect to the translations in R^n. Sometimes one regards the analogous but technically simpler case of a discretisized Schrödinger equation (Jacobi matrix) with ergodic potential V in $l^\infty(Z^n)$:

$(H_\omega \varphi)(x) := -\sum_{|y-x|=1} \varphi(y) + (2n + V_\omega(x))\,\varphi(x)$. Ergodicity of V is now with respect to translations in Z^n. The one-dimensional case describes via the transformation $x \to t$ (time) vibrating systems with ergodic varying: $y'' = V_\omega(t)\,y(t)$; this is connected with the notion "Parametric resonance". (For references, precise definitions and results cf. [1]).

The hull h of V_ω is defined as the closure of $\{V_\omega(x-x_0)\,;\, x_0 \in R^n\}$

---

1) Sektion Mathematik, Karl-Marx-Universität, DDR-7010 Leipzig

in a certain topology of L^p_{loc}. Because of ergodicity, for almost all ω $V_\omega(x)$ is recurrent and the hull $h(V_\omega)$ is independent on ω, such that one can prove the following theorem:

Theorem 1: For almost all ω, ω' : $\sigma_{disc} = \emptyset$ and $\sigma(H_\omega) = (H_{\omega'})$.

Pastur proved (with the help of the ergodic properties of the dimension of the spectral projectors) that every fixed λ is with probability 0 an eigenvalue of H_ω with finite multiplicity. Kirsch/Martinelli found that the singular, absolutely continuous and the point spectrum is a set independent on ω for a.e. ω.

We believe that the following problem is the key-problem in spectral analysis of ergodic operators:
Which information one can derive for the classification of the spectra $\sigma(H_\omega)$ from the classification of the ergodic processes V_ω?

In the following we want to refer the known results; in parentheses we mention the year of publishing the first results.

3) $V \equiv 0$
Then $\sigma = \sigma_{ac} = [0, \infty)$.

4) V_ω are periodic.
Then $\sigma = \sigma_{ac} = \bigcup_{i=1}^{\infty} [a_i, b_i]$ (1974). These potentials describe periodic crystals.

5) V_ω are almost periodic, n=1.
These potentials occur in solid state physics, connected with the notions "Incommensurate lattice" and "Devil's stair case".
Avron/Simon tried to explain the structure of the rings of saturn by an almost periodic Schrödinger equation. It is remarkable that Cantor sets should have physical relevance.

5.1) Finite zone potentials.
Dubrovin e.a. (1976) formulated a condition (such a condition exists also for periodic potentials) such that $\sigma = \sigma_{ac} = \bigcup_{i=1}^{N} [a_i, b_i] \cup [a_{N+1}, \infty)$.
For all types of spectra in 1) - 5.1) Avron/Simon proposed the notion of "ordinary" spectrum. The next examples show all "extra-ordinary" behaviour:

5.2) V_ω are limit-periodic.
Moser, Avron/Simon and Chulaevskij (1981) constructed independently a large class of limit periodic potentials such that $\sigma = \sigma_{ac}$ is a Cantor set with positive Lebesgue measure. Avron/Simon called such

extraordinary absolutely continuous spectra recurrent; the ordinary absolutely continuous spectra are called transient.

5.3) V_ω are quasiperiodic.

Belokolos and Dinaburg/Sinai (1975) proved for a general class the occurrence of an absolutely continuous part in the spectrum, which is at least a Cantor set.

Avron/Simon (1981) proved for special examples of Jacobi matrices with cosine-potential and $\tilde{\lambda} > 2$ that $\sigma = \sigma_{sc}$, i.e. the spectrum is singular continuous. (Pearsson (1978) gave the first example of Schrödinger equations with $\sigma = \sigma_{sc}$, though for a nonergodic potential.) Avron/Simon believe that for the general case of discretisized Schrödinger equations with cosine potential and $\tilde{\lambda} > 2$ only pure point spectrum σ_p occurs. It is a conjecture of André/Aubry that the spectral class can change in dependence on the coupling constant. That is why one cannot expect to get necessary and sufficient conditions in terms of the invariants of the ergodic processes in order to classify the spectrum of H_ω.

6) V_ω is "purely random", n=1.

Goldsheid/Molchanov/Pastur (1977) constructed potentials by a diffusion process on a compact manifold M and a function F on M via $V_\omega(x) := F(\omega(x))$ and proved that for almost all ω $\sigma(H_\omega) = \sigma_p = [\inf F, \infty)$. Perhaps this result is true for arbitrary weakly mixing processes. Molchanov (1981) announced a proof that for $n \geq 5$ the Jacobi matrices with random, independent potential in every point of Z^n have only absolutely continuous spectrum.

Physicists believe that for some n purely random potentials cause a "mobility edge" E, i.e. $\sigma_{(-\infty, E]} = \sigma_{ac}$ and $\sigma_{[E,\infty)} = \sigma_p$. A nonrandom Hamiltonian with such a behaviour of the spectrum was given by Miller/Simon.

Examples 5) and 6) are not only examples of ergodic potentials; the proofs for the results concerning the spectrum also use ergodic theory, e.g. Oseledets' and the subadditive ergodic theorem and a theorem of Furstenberg type.

7) V_ω describes substitutional alloys.

Let ω be a random process with values in N^{Z^n} and define $V_\omega(x) := \sum_{i \in Z^n} V^{\omega(i)}(x-i)$, where $\{V^j\}$ is a set of potentials in $L^p(R^n)$ with a common L^p-majorant. (V^j is the potential for the jth atomic sort; the process ω describes the random occupation of lattice points in Z^n by atoms.) Englisch/Kürsten [1] proved for a general class of random processes (i.e. for such processes, that for almost every ω $h(\omega) = N^{Z^n}$) that for a.e. ω

$\sigma(H_\omega) = \bigcup \sigma(H_{per}) = \bigcup_{i=1}^{\infty} [a_i, b_i]$, where H_{per} denotes arbitrary periodic Hamiltonians from $\{H_\omega; \omega \in N^{Z^n}\}$ and the union is taken over all this periodic Hamiltonians. A related result was proved independently by Kirsch/Martinelli (1980). This result implies that the concept of bowing in solid state physics is wrong, i.e. that the widths of the gaps (b_i, a_{i+1}) should depend nonlinear on the concentration of the atomic sorts. If all concentrations are nonzero, the widths are constant! This means that not the width of the gaps, but the density of states is the physical object which one can study in dependence on the atomic concentration. While this result is true for any space dimension, the next one is proved only for $n = 1$:

If all V^j have the interval $[-1/2, +1/2]$ as support and are symmetric with respect to $x = 0$, then for the above mentioned class of processes and a.e. ω $\inf \sigma(H_\omega) = \inf_j \inf \sigma(-d^2/dx^2 + \sum_{i \in Z} V^j(x-i))$. For nonsymmetric V^j there are counterexamples.

Generalizing and correcting the results of Luttinger (1951) and Dworin (1965) we gave in [1] further characterizations of $\sigma(H_\omega)$ as set. We conjecture that for a.e. ω, $\sigma(H_\omega) = \sigma_p$. Molchanov/Seidel (1982?) announce that they can apply the methods of Goldsheid/Molchanov/Pastur (cf. example 6) to jump processes ω; perhaps this contains a proof of the conjecture. If ω is a Markov process, then a result of Royer shows, that $\sigma(H_\omega)$ has with probability 1 no absolutely continuous part in the spectrum.

References.

[1] Englisch,H., Kürsten,K.D.: The spectrum of ergodic Schrödinger operators. Preprint in preparation.

[2] Reed,M., Simon,B.: Methods of Modern Mathematical Physics. 4 (1978), 1st edition, New York e.a.

ON THE SPATIAL CONSTANT OF SUPERADDITIVE SET FUNCTIONS IN $\mathbb{R}^d$

Christian Grillenberger[1] and Ulrich Krengel[1]

Abstract: The limit property for parallelepipeds defining the spatial constant of superadditive translation invariant set functions remains true for convex sets, but not for general van-Hove sequences.

Introduction: The spatial constant of a subadditive (translation invariant) set function g is the limit of the "average value" of g for sequences of parallelepipeds whose width tends to infinity. Extending a result of Robinson and Ruelle [5], Nguyen [3] showed that the parallelepipeds may be replaced by sequences of general convex sets, if g is even strongly subadditive. His proof carries over to van-Hove sequences, i.e., sequences of Borel sets whose boundary is asymptotically small compared to the volume. We show that the strong subadditivity is not needed for convex sets, but it is essential for general van-Hove sequences. The crucial property of convex sets needed in our proof is the existence of parallelepipeds containing them and filled by them up to a constant factor (1/d!). For technical reasons and also for reasons of personal taste, we use superadditive instead of subadditive set functions. Since no positivity is required we may do this by a change of signs.

$\mathcal{B}$ denotes the family of bounded Borel sets B in $\mathbb{R}^d$ with strictly positive Lebesgue measure $\lambda(B)$. $\mathcal{K}$ is the family of convex sets in $\mathcal{B}$. Put $U_r(x) := \{y \in \mathbb{R}^d : \|x-y\| < r\}$, where $\|\cdot\|$ is the Euclidean metric, and

$$\rho(B) := \sup\{r \geq 0 : \exists\, x \in B \text{ with } U_r(x) \subset B\},$$

$$\partial_h B := \{y \in \mathbb{R}^d : \exists\, x \in \partial B \text{ with } \|y-x\| < h\} = \bigcup_{y \in \partial B} U_h(y).$$

A parallelepiped P is an affine image of a half open cube $W_k = \{y=(y_1,\dots,y_d) \in \mathbb{R}^d : 0 \leq y_i < k\}$ belonging to $\mathcal{B}$. We always work with half open parallelepipeds and cubes, so that $\mathbb{R}^d$ has a covering by disjoint translates of them. A sequence (B_k) in $\mathcal{B}$ is called a van-Hove sequence (or van-Hove convergent to $\mathbb{R}^d$) if $\lambda(\partial_h B_k)/\lambda(B_k)$ tends to 0 for $k \to \infty$ for all $h > 0$. This implies $\lambda(B_k) \to \infty$. A set function $g : \mathcal{B} \to \mathbb{R}$ is called translation invariant, if $g(B+u) = g(B)$ holds for all $B \in \mathcal{B}$ and all $u \in \mathbb{R}^d$. g is called superadditive if

$$g(B_1 \cup B_2) \geq g(B_1) + g(B_2)$$

holds for all disjoint $B_1, B_2 \in \mathcal{B}$, and g is called strongly superadditive if, in addition

$$g(B_1 \cup B_2) + g(B_1 \cap B_2) \geq g(B_1) + g(B_2)$$

1) Institut für Mathem. Statistik, Univ. Göttingen, Lotzestr. 13, D-3400 Göttingen

holds for all $B_1, B_2 \in \mathcal{B}$ with $B_1 \cap B_2 \in \mathcal{B}$. The spatial constant of a translation invariant superadditive g is defined by

$$\gamma := \gamma(g) := \lim_{k\to\infty} g(W_k)/\lambda(W_k).$$

It is easy to see that the limit exists, and that $\gamma(g) = \sup\{g(W_k)/\lambda(W_k) : k > 0\}$.

It is also simple to check that in both expressions for $\gamma(g)$ the sets W_k may be replaced by parallelepipeds of the form $(\{y \in \mathbb{R}^d : 0 \le y_i < a_i(k)\}$ with $a_i(k) \to \infty$ $(k\to\infty)$ for $i=1,\dots,d$. Put

$$i(g) = \inf\{g(B) : B \in \mathcal{B} \text{ and } B \subset W_1\}$$

This note has been motivated by the following theorem of Nguyen X.X.:

Theorem 1:

If g is translation invariant and strongly superadditive with $i(g) > -\infty$, and if (H_n) is a von Hove sequence, then $g(H_n)/\lambda(H_n)$ tends to γ.

In [3,prop.2.14] this result is formulated only for sequences H_n in $\mathcal{K}$ with $\rho(H_n) \to \infty$, but it is mentioned that the proof works for general van Hove sequences. The special case $g \ge 0$ is due to Robinson-Ruelle [5]. Below we shall show by example that theorem 1 does not remain true for all van-Hove sequences when the translation invariant g with $i(g) > -\infty$ is only superadditive. However, for sequences in $\mathcal{K}$ we have:

Theorem 2:

If g is translation invariant and superadditive with $i(g) > -\infty$, and if (K_n) is a sequence in $\mathcal{K}$ with $\rho(K_n) \to \infty$, then $g(K_n)/\lambda(K_n)$ tends to γ.

Lemma 1: [2]

For any $K \in \mathcal{K}$ there exists a parallelepiped P with $\lambda(P) \le d!\lambda(K)$, the closure of which contains K.

Clearly, the lemma implies that for any $K \in \mathcal{K}$ and for any $c_d > d!$ there exists a parallelepiped P with $P \subset K$ and $\lambda(P) < c_d\, \lambda(K)$.

Lemma 2:

Any sequence (K_n) in $\mathcal{K}$ with $\rho(K_n) \to \infty$ is a van-Hove sequence.

This is stated in [4]. A simple proof may be found in [1].

Proof of theorem 2:

We delete the proof of the inequality

$$\liminf g(K_n)/\lambda(K_n) \ge \gamma.$$

It is an easy consequence of lemma 2 if g is nonnegative. In the general case some additional estimates are needed, but they are similar to

those used below.
Now assume

$$\limsup_{m\to\infty} g(K_m)/\lambda(K_m) > \gamma.$$

Fix $c_d > d!$, and choose $0 < \varepsilon < 1/4$ with

$$\varepsilon(|\gamma| + |i(g)|) < 1 \quad \text{and} \quad \limsup_{m\to\infty} g(K_m)/\lambda(K_m) > \gamma + 4\varepsilon c_d, \tag{1}$$

and $k \in \mathbb{N}$ with

$$g(W_k)/\lambda(W_k) > \gamma - \varepsilon. \tag{2}$$

By Lemma 2 we may find an $n \in \mathbb{N}$ with

$$g(K_n)/\lambda(K_n) > \gamma + 4\varepsilon c_d \quad \text{and} \quad \lambda(\partial_{k\sqrt{d}} K_n)/\lambda(K_n) < \varepsilon^2. \tag{3}$$

Lemma 1 and the choice of c_d imply the existence of a parallelepiped P with $K \subset P$ and $\lambda(P) \le c_d\lambda(K_n)$. Let $\delta = \{\|x-y\| : x,y \in P\}$ denote the diameter of P. If $u \in \mathbb{N}$ is large enough we have

$$\lambda(\partial_{2\delta} W_u)/\lambda(W_u) < \varepsilon. \tag{4}$$

Let $A \subset \mathbb{R}^d$ be a countable lattice for which the sets $(a+P)$ with $a \in A$ form a disjoint covering of $\mathbb{R}^d$. Put

$$A_1 := \{a \in A : (P+a) \cap (W_u \setminus \partial_\delta W_u) \neq \emptyset\},$$

$$A_2 := \{a \in A : (P+a) \cap \partial_\delta W_u \neq \emptyset\}.$$

Let $C = \{k\cdot b : b \in \mathbb{Z}^d\}$, so that $\{c+W_k : c \in C\}$ is a disjoint covering of $\mathbb{R}^d$, and let

$$C_1 := \{c \in C : c + W_k \subset \bigcup_{a\in A_1} (K_n+a)\},$$

$$C_2 := \{c \in C \setminus C_1 : (c+W_k) \cap (\bigcup_{a\in A_1} (K_n+a)) \neq \emptyset\},$$

$$C_3 := \{c \in C \setminus (C_1 \cup C_2) : c + W_k \subset W_u\}.$$

We may assume that u is a multiple of k, so that W_u is the disjoint union of the sets W_k+c with $c \in C_1 \cup C_2 \cup C_3$. By the definition of A_1 and A_2

$$W_u \setminus \partial_\delta W_u \subset \bigcup_{a\in A_1} (a+P) \subset W_u \subset \bigcup_{a\in A_1\cup A_2} (a+P).$$

If $|A_2|$ denotes the number of elements of A_2, (4) yields

$$\lambda(\bigcup_{a\in A_2}(a+P)) = |A_2|\lambda(P) \le \lambda(\partial_{2\delta}W_u) \le \varepsilon\lambda(W_u).$$

Now

$$g(W_u) \ge g(\bigcup_{c\in C_1\cup C_2}(c+W_k)) + \sum_{c\in C_3} g(c+W_k)$$

$$\ge \sum_{a\in A_1} g(a+K_n) + \sum_{c\in C_2} g((c+W_k) \setminus \bigcup_{a\in A_1}(a+K_n)) + |C_3| g(W_k)$$

$$\ge |A_1|(\gamma+4\varepsilon c_d)\lambda(K_n) - |C_2|\,|i(g)|\,\lambda(W_k) + |C_3|\,(\gamma-\varepsilon)\lambda(W_k).$$

Using $W_u = \bigcup_{a\in A_1}(a+K_n) \cup \bigcup_{c\in C_2\cup C_3}(c+W_k)$, we may continue

$$g(W_u) \ge \gamma\cdot\lambda(W_u) + |A_1|4\varepsilon c_d\lambda(K_n) - |C_2|(|\gamma| + |i(g)|)\lambda(W_k) -$$

$$- \varepsilon|C_3|\,\lambda(W_k).$$

As all sets $c+W_k$ with $c \in C_2$ are contained in $\partial_{k\sqrt{d}}(a+K_n)$ for some $a \in A_1$ we get

$$g(W_u) \ge \gamma\,\lambda(W_u) + 4\varepsilon|A_1|\,\lambda(P) - \varepsilon^{-1}|A_1|\lambda(\partial_{k\sqrt{d}}K_n) - \varepsilon\lambda(W_u)$$

$$\ge \gamma\,\lambda(W_u) + 4\varepsilon\lambda(W_u)(1-\varepsilon) - \varepsilon|A_1|\lambda(K_n) - \varepsilon\,\lambda(W_u)$$

$$\ge \gamma\,\lambda(W_u) + \varepsilon\,\lambda(W_u).$$

As u was arbitrarily large this means that $\limsup_{i\to\infty} g(W_i)/\lambda W_i) \ge \gamma+\varepsilon$, a contradiction. □

<u>The counterexample</u>:

Let $d = 1$. We now construct a nonnegative superadditive set function $g : \mathcal{B} \to \mathbb{R}$ and a van-Hove sequence with

$$\gamma(g) < 1 \quad\text{and}\quad \lim g(H_n)/\lambda(H_n) = 1.$$

For $n \ge 1$ let $h_n = 2^{2^n}$. Then $h_n = h_{n-1}^2$. For $n \ge 2$ let

$$H_n = [-h_n, 0[\,\cup \bigcup_{i=1}^{h_{n-1}} [ih_n - h_{n-1}, ih_n[.$$

$I(H_n)$ denotes the smallest interval containing H_n. It is easy to see that (H_n) is a van-Hove sequence. H_n consists of one interval of length h_n and h_{n-1} intervals of length h_{n-1}; the gaps in between have length $h_n - h_{n-1} < h_n$, they are shorter than the short intervals of H_{n+1}. Thus, if $a+H_m$ and $b+H_n$ are disjoint, $m < n$, and $a + I(H_m)$ intersects $b+I(H_n)$, then $a+H_m$ is contained in one of the gaps of $b+H_n$. Similarly, $a+H_n$ and $b+H_n$ can be disjoint only if $a+I(H_n)$ and

$b+I(H_n)$ are disjoint.
Let

$$g(B) = \sup\{\lambda(H): H\subset B,\ H = \bigcup_{n=1}^{\infty} \bigcup_{j=1}^{k_n} (a_{j,n}+H_n) \quad \text{(disjoint union)}\}.$$

It is clear that g is translation invariant and superadditive, and $g(H_n) = \lambda(H_n)$. We only have to show $\gamma(g) < 1$. We have

$$\lambda(H_n)/\lambda(I(H_n)) = 2h_n/(1+h_{n-1})h_n < 2/h_{n-1}.$$

Together with the disjointness properties mentioned above this shows that for each <u>interval</u> B any disjoint union H of translates of sets H_n, contained in B, can cover B only up to a rest of size

$$\lambda(B) \cdot \prod_{n\geq 2} (1-2/h_{n-1}).$$

Hence

$$\gamma(B) \leq 1 - \prod_{n\geq 1} (1-2/h_n) < 1.$$

The last inequality holds because all factors are positive and $\sum 2/h_n \leq \sum 2^{-n} < \infty$.
Of course, similar examples exist in every $\mathbb{R}^d$.
The result and example here were given for set functions on $\mathcal{B}$. A similar result holds if $\mathbb{R}^d$ is replaced by $\mathbb{Z}^d$, $\mathcal{B}$ by $\mathcal{B}\cap\mathbb{Z}^d$, $\mathcal{K}$ by $\mathcal{K}\cap\mathbb{Z}^d$, and λ by the counting measure. In this case $i(g) > -\infty$ is automatically satisfied.

<u>References</u>:

[1] U.KRENGEL: Ergodic Theorems. de Gruyter,Berlin; to appear.

[2] H.HADWIGER: Volumschätzung für die einen Eikörper überdeckenden und unterdeckenden Parallelotope. Elem.Math. <u>10</u>, 122-124,(1955).

[3] X.X.NGUYEN: Ergodic theorems for subadditive spatial processes. Z.Wahrscheinlichkeitstheorie verw.Geb. <u>48</u>, 159-176,(1979).

[4] X.X.NGUYEN and H.ZESSIN: Punktprozesse mit Wechselwirkung.Z.Wahrscheinlichkeitstheorie verw.Geb. <u>37</u>, 91-126,(1976).

[5] D.W.ROBINSON and D.RUELLE: Mean entropy of states in classical statistical mechanics. Comm.Math.Phys. <u>5</u>, 288-300,(1967).

ADDITIVE INTEGRALS OF MOTION OF POINT PARTICLE SYSTEMS IN ONE DIMENSION

B. M. Gurevich[1)]

The problem of additive integrals of motion of a finite number of interacting point particles is related to the problem of invariant measure for infinite-dimensional dynamical systems.

Let M be the phase space of an infinite point particle system in R^d, $d \geq 1$, i.e. the family of locally-finite sets $X \subset R^d \times R^d$ consisting of pairs $x = (q,p)$, where q is the position and p the momentum of a particle $x \in X$.

Let H^o denote a formal Hamiltonian describing the particle interaction. Assume that H^o is of the form

$$H^o(X) = \sum_{(q,p)\in X} \frac{1}{2}p^2 + \sum_{(q,p),(q',p')\in X} U(|q-q'|), \quad X\in M, \tag{1}$$

where U is a pair potential such that $U(r) \longrightarrow 0$ as $r \longrightarrow \infty$.

There are some works where an infinite-dimensional dynamical system corresponding to H^o is constructed (see [3] for a review of most of these works).

A probability measure on M can be given by its correlation measures [3,4] and, under some additional regularity conditions, by its correlation functions. If a measure m on M is invariant under the particle dynamics, one usually succeeds in proving (under some restrictions on m and U) that its correlation functions are a stationary solution of the Bogoliubov hierarchy equations [1] (or the BBGKY hierarchy).

A natural class of probability measures on M is that of Gibbsian measures (or DLR-measures [2,7]). By the definition of such a measure one can associate with it a formal Hamiltonian H of the form

$$H(X) = \sum_{k=0}^{\infty} \sum_{Y\in M_k(X)} f_k(Y), \quad X\in M. \tag{2}$$

Here $M_k(X)$ is the set of k-point subsets of X, and f_k, $k \geq 1$, a real-valued function on M_k, the phase space of a k-particle system, satisfying

1) Moscow State University, 117234 Moscow, USSR.

$$\lim_{|q|\to\infty} f_k((q,p)\cup X) = 0, \quad k \geqslant 2, \quad X \in M_{k-1}, \quad p \in R^d. \qquad (3)$$

(One lets $f_0 = 0$). Assumption (3) garantees the uniqueness of representation (2) provided the right side of (2) is well-defined as a function on M.

Let $\bar{H}_n$ denote the set of formal Hamiltonians H (see (2)) for which $f_k = 0$ when $k > n$ and let $\bar{H}_{fin} = \bigcup_k \bar{H}_k$. Evidently, $H^o \in \bar{H}_2$.

In [4] those Gibbsian measures are described whose formal Hamiltonians belong to $\bar{H}_{fin}$ and whose correlation functions are a stationary solution of the Bogoliubov hierarchy equations corresponding to H^o. It is shown that if a formal Hamiltonian $H \in \bar{H}_n$, $n \geqslant 2$, is associated with such a measure, then the functions f_k in (2) satisfy the following equations:

$$\{f_k(X), H_k^o(X)\} + \sum_{x=(q,p)\in X} \{f_{k-1}(X\setminus x), \sum_{y=(q',p')\in X\setminus x} U(|q-q'|)\} = 0, \quad X \in M_k, \ k \geqslant 1, \qquad (4)$$

where $\{\ \}$ are the Poisson braces and H_k^o is the restriction of H^o to M_k.

Notice that (4) is the conjugate system of the Bogoliubov hierarchy with respect to the inner product defined by

$$\langle F, G \rangle = \sum_i \int_{M_i} f_i g_i \, dl_i,$$

where $F = (f_i)_{i=1}^{\infty}$, $G = (g_i)_{i=1}^{\infty}$, $f_i, g_i : M_i \to R^1$, and l_i is the measure on M_i obtained from Lebesgue measure by symmetrization.

Let H_k, $k \geqslant 1$, denote the restriction of H to M_k. This restriction is well-defined as a function on M_k. It can be easily deduced from (4) that for every $k \geqslant 1$, H_k is a first integral of the Hamiltonian system with Hamiltonian H_k^o. Motivated by this fact we give the following definition.

<u>Definition.</u> (cf. [6]). A function $h: M_n \to R^1$, $n \geqslant 1$, is called an additive first integral of the Hamiltonian system with Hamiltonian H_n^o (shortly the H_n^o-system) if there exist functions $g_k : M_k \to R^1$, $1 \leqslant k \leqslant n$, such that: (a) $g_n = 0$ and g_k satisfies (3) when $k \geqslant 2$,

(b) for $k \leq n$ the function

$$h_k(X) = \sum_{i=1}^{k} \overline{\sum_{Y \in M_i(X)}} g_i(Y), \quad X \in M_k ,$$

is a first integral of the H_k^o-system.

From the above it follows that one can associate an additive first integral of the H_n^o-system with every Gibbsian measure whose formal Hamiltonian H belongs to $\overline{H}_{n-1}$ and whose correlation functions are a stationary solution of the Bogoliubov hierarchy (this integral is H_n).

In [4] we have described all such inegrals satisfying some additional regularity conditions. It turned out that they were exhausted by the linear combinations of the classical first integrals: the particle number, the energy, and the momentum and angular momentum components (when $d > 1$). From this it follows that among the Gibbsian measures with Hamiltonians in $\overline{H}_{fin}$ only the equilibrium Gibbsian measures with Hamiltonian H^o can be invariant under the infinite-dimensional dynamical system induced by H^o. This result (see [5] for its proof in the simplest case where $n = 3$) is obtained for a finite range potential only. Despite the fact that the proof is essentially based on this assumption, one could think that it was only technical and that it could be replaced by a sufficiently fast decrease at infinity. But there are examples showing that this is not the case. Among them the most suitable for statistical mechanics is the potential $U^o(r) = 1/\mathrm{sh}^2 r$ which is a special case of the Weierstrass elliptic function $\wp$. By the L,A-pair method one can find the following first integral of the H_n^o-system with $U = U^o$, $n \geq 4$, $d = 1$:

$$h(X) = h(q_1, p_1, \dots, q_n, p_n) = \sum_{i=1}^{n} p_i^4 + \overline{\sum_{i,j:i=j}} (p_i^2 + p_i p_j + p_j^2) U(|q_i - q_j|) +$$

$$+ \overline{\sum_{\substack{i_1, i_2, i_3, i_4: \\ i_1 = i_2 = i_3 = i_4}}} U(|q_1 - q_2|) U(|q_2 - q_3|) U(|q_3 - q_4|) U(|q_4 - q_1|), \quad X \in M_n. \qquad (5)$$

This function is not a linear combination of the classical first integrals and when $n \geq 5$ (in fact when $n = 4$ as well) it satisfies all the conditions from the above definition of the additive first integral. As is shown by V.A.Chulaevski, there exists a Gibbsian measure m on M with a formal Hamiltonian $H \in \bar{H}_4$ such that H_n, the restriction of H to M_n, $n \geq 4$, coincides with (5). Moreover, m is invariant under the infinite-dimensional dynamics induced by H^o, where $U = U^o$ (m is certainly non-equilibrium with respect to H^o).

We now state the main results of the present paper which will show that the examples like this can be regarded as an exception.

Theorem 1. Let $U \in C^\infty(0,\infty)$, $\lim\limits_{r \to \infty} U(r) = 0$, $\lim\limits_{r \to 0} U(r) = +\infty$ and assume that for some $n \geq 4$ the H_n^o-system in one dimension has an additive first integral $h \in C^\infty$ which is not a linear combination of the classical first integrals. If, moreover, either

(a) $\int_{r_o}^{\infty} r^k U(r)dr < \infty$ for some $r_o > 0$ and every $k \geq 1$, or (b) h is translation invariant with respect to the particle positions when the particle momenta are fixed, then

$$U'(s)[U(t) - U(s+t)] + U'(t)[U(s+t) - U(s)] + \\ + U'(s+t)[U(t)-U(s)] = 0, \quad s,t > 0. \tag{6}$$

Remark. In the proof of this theorem we actually only need that $U \in C^k$, $h \in C^l$ for some k,l depending on n.

Equation (6) occurs in the theory of elliptic functions. It is satisfied by the Weierstrass function $\wp$ (see, for example,[10]). It turns out that (6) determines this function to within a linear substitution.

Theorem 2. Every solution of (6) of class C^3 has the form

$$c_1 + c_2 \wp,$$

where c_1, c_2 are constants.

Remarks. 1. Mrs. I.Ju.Musina previously proved this assertion for a more restricted class of functions U .

2. Theorem 2 implies that there is only a finite parameter class of potentials vanishing at infinity for which there exist additive first integrals other than those in the case of a finite range potential. It is worth noting that for every of these potentials the corresponding Hamiltonian system is completely integrable [8,9].

References

[1] Bogoliubov,N.N.: Problems of the Dynamical Theory in Statistical Physics. Moscow, 1946.

[2] Dobrushin,R.L.: Gibbsian Random Fields. The General Case. Funkts. Analiz. 3/1 (1969).

[3] Gurevich,B.M., Oseledec,V.I.: Some Mathematical Problems Related to the Non-Equilibrium Statistical Mechanics of Infinite Many Particles. Probab. Theory, Math. Statist., Theor. Cybern. The Institute for Scientif. Inform. 14 (1977).

[4] Gurevich,B.M., Suhov,Ju.M.: Stationary Solutions of the Bogoliubov Hierarchy Equations 1 . Comm. Math. Phys. 49 (1976), 63-96.

[5] Gurevich,B.M., Suhov,Ju.M.: Stationary Solutions of the Bogoliubov Hierarchy Equations 2 . Comm. Math. Phys. 54 (1977), 81-96.

[6] Gurevich,B.M., Suhov,Ju.M.: Additive Integrals of Motion of Pairwise Interacting Particles. Uspehi Mat. Nauk 36/4 (1981).

[7] Lanford,O.E., Ruelle,D.: Observations at Infinity and States with Short Range Correlations in Statistical Mechanics. Comm. Math. Phys. 13 (1969), 194-215.

[8] Olshanetsky,M.A., Perelomov,A.M.: Classical Integrable Finite-Dimensional Systems Related to Lie Algebras. Phys. Reports 71/5 (1981).

[9] Pidkuyko,S.I.: On the One-Dimensional Integrable N Body Problem. Uspehi Mat. Nauk 33/3 (1978).

[10] Whittaker,E.T., Watson,G.N.: A Course of Modern Analysis. Cambridge 1948.

BORDERING FACTORS OF DYNAMICAL SYSTEMS

K. HÄSLER [1]), H. MICHEL [1]) and G. SCHWENZFEGER [2])

In a previous paper [7] the second of the authors introduced a notion of support for dynamical systems by simulation of the fact that to every natural number n there corresponds the set $\pi(n)$ of those primes which occur in the standard form of n.

In this paper we shall deal with notions of partial maximality ("bordering") of factors of dynamical systems. At first the counterparts to these notions in the set of natural numbers are explained. If $\varepsilon(n,p)$ is the exponent of $p \in \pi(n)$, we define for a divisor n' of n and the set $P(n',n) = \{p \in \pi(n) | \ \varepsilon(n',p) = \varepsilon(n,p)\}$

$$n' \text{ is n-bordering} \Leftrightarrow |P(n',n)| > 0,$$

$$n' \text{ is narrowly n-bordering} \Leftrightarrow |P(n',n)| = 1,$$

where $|.|$ denotes the number of elements.

For at least two reasons the situation for dynamical systems is much more complicated than in the set of natural numbers. Interesting dynamical systems do not have an equivalent of the standard form and the circumstances are not so "discrete" in general as in the number case. Therefore our motivating background in section 2 works with functions instead of numbers.

Some of the following results should be mentioned. If D is a totally ergodic dynamical system with discrete spectrum and a proper value group H(D) that is generated by a finite set of rationally independent

1) Martin-Luther-Universität Halle-Wittenberg, Sektion Mathematik, Universitätsplatz 6, DDR-4010 Halle(S.)

2) Pädagogische Hochschule Halle N. K. Krupskaja, Sektion Mathematik/Physik, Burgstr. 46, DDR-4020 Halle(S.)

numbers $c_1,\dots,c_k$, then any group H', generated by an arbitrary subset of $\{c_1,\dots,c_k\}$, is the proper value group of a bordering factor of D, but there are more bordering factors than these ones (theorem 3.1). If D has positive entropy then D is Bernoulli iff D itself is the only D-bordering factor (mod isom), [see theorem 3.7.] In the class of all narrowly bordering factors of D exists an equivalence relation that simulates in some sense the prime number représentation of natural numbers. The equivalence classes are certain analoga of the prime powers (sect. 4, infra).

1. Notations and definitions

In the following a dynamical system is a quadruple $D=(X,\mathcal{F},m,T)$, where $(X,\mathcal{F},m)$ is a Lebesgue measure space and T a measurable measure-preserving transformation from X into itself. Together with T the composition T^n is endowed with the same properties. Therefore

$$D^n = (X,\mathcal{F},m,T^n) \tag{1}$$

is well defined. $D_1=(X_1,\mathcal{F}_1,m_1,T_1)$ is called a factor of the system $D_2=(X_2,\mathcal{F}_2,m_2,T_2)$, if there exist sets $\tilde{X}_1,\tilde{X}_2$ and a mapping $S:\tilde{X}_2\to\tilde{X}_1$ with $S^{-1}(\tilde{\mathcal{F}}_1)\subset\tilde{\mathcal{F}}_2$, $m_2(S^{-1}A)=m_1(A)$ $(A\in\tilde{\mathcal{F}}_1)$, $m_2(T_2^{-1}S^{-1}A\triangle S^{-1}T_1^{-1}A)=0(A\in\tilde{\mathcal{F}}_1)$, where $\tilde{X}_i$ are measurable subsets of X_i such that $m_i(\tilde{X}_i\triangle X_i)=0$ and $\tilde{\mathcal{F}}_i=\tilde{X}_i\cap\mathcal{F}_i$ $(i=1,2)$. In this case we write $D_1\overset{S}{\leq}D_2$ or shortly, if S is not essential, $D_1\leq D_2$. This means the same as to define

$$D_1\leq D_2 \iff \exists S:\ D_1\overset{S}{\leq}D_2.$$

Then this is a preorder $(D_1\leq D_1;\ D_1\leq D_2, D_2\leq D_3\Rightarrow D_1\leq D_3)$, where S=id and $S_{13}=S_{12}\circ S_{23}$ are available in the axioms of reflexivity and transitivity (resp.). Furthermore the implication

$$D_1\leq D_2 \Rightarrow D_1^n\leq D_2^n\ (n\in N) \tag{2}$$

is valid and a suitable $S:\tilde{X}_2\to\tilde{X}_1$ on the left side is available on the right as well (independent of n).

Two kinds of equivalence are used. D_1,D_2 are called weakly isomorphic (Ja.G.Sinai [11]), if $D_1\overset{S_1}{\leq}D_2$ and $D_2\overset{S_2}{\leq}D_1$ for some possibly different mappings $S_1:\tilde{X}_2\to\tilde{X}_1$, $S_2:\tilde{X}_1\to\tilde{X}_2$ as defined above. D_1,D_2 are called isomorphic, if $D_1 \overset{S}{\leq} D_2$ with a factor mapping S which is invertible mod 0. There is a result showing these two equivalence relations to be different (J.Clark [1]).

2. Bordering factors

In order to settle a notion of bordering for dynamical systems we consider at first a similar situation for real functions. Let $f,g:(0,\infty)\to R$ be two increasing functions with the properties

$$x \leq g(x) \leq f(x) \quad (x\in(0,\infty)). \qquad (3)$$

At least in case that f and g have suitable smoothness properties it is easy to check whether there is one

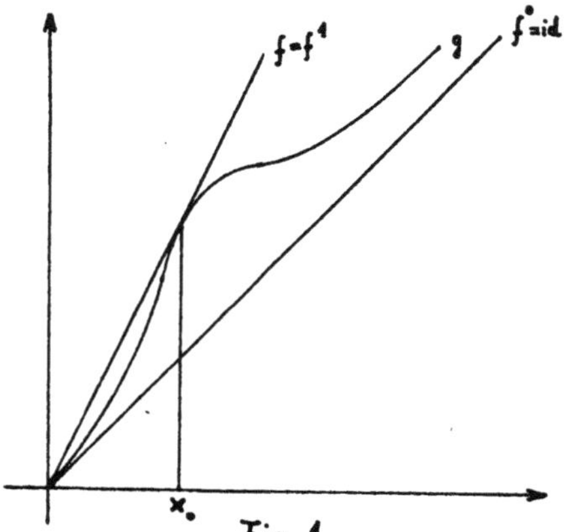

Fig. 1

or several points $x_0 \in (0,\infty)$ with $g(x_0)=f(x_0)$. Together with (3) this means that the graph of g borders the graph of f from below. Of course, usual techniques of calculus to prove this bordering property break down, if f and g are replaced by dynamical systems D and D_0 and the relation $g\leq f$ by the factor relation $D_0\leq D$. But there is another possibility to transfer the notion of bordering: embed f in a flow $(f^s)_{s\in R}$, $f^1=f, f^0=id, f^s\circ f^t=f^{s+t}, (s,t\in R)$ and express the bordering property of g by

$$g\leq f^s \Longrightarrow 1\leq s. \qquad (4)$$

Now all terms in (4) can be made meaningful for dynamical systems as well, but even now it is not always possible to embed a dynamical system in a real flow (e.g. for suitable mixing transformations [2], for suitable K-automorphisms [10], for totally ergodic dynamical systems with discrete spectrum [5], [8]). Therefore we avoid the full parameter group R and possible difficulties with roots and demand in the function case

$$g^m \leq f^n \Rightarrow m \leq n \quad (m,n \in N).$$

Stimulated by these considerations we pass to dynamical systems.

2.1. Definition. A factor D_o of a dynamical system D is called D-bordering from below if

$$(m,n \in N \text{ and } D_o^m \leq D^n) \Rightarrow m \leq n, \tag{5}$$

resp. D-bordering from above if

$$(m,n \in N \text{ and } D_o^m \leq D^n) \Rightarrow m \geq n. \tag{6}$$

The class of all D-bordering factors from below (resp. above) is denoted by $\underline{\mathfrak{B}}(D)$ (resp. $\overline{\mathfrak{B}}(D)$).

We discuss these notions not referring to functions.

2.2. Remarks. 1. If $\underline{\mathfrak{B}}(D) \neq \emptyset$, then $D \in \underline{\mathfrak{B}}(D)$.
Let $D_o \in \underline{\mathfrak{B}}(D)$. $D \notin \underline{\mathfrak{B}}(D)$ would imply $D^{m_1} \leq D^{n_1}$ and $m_1 > n_1$ for some $m_1, n_1 \in N$. From (1),(2) and $D_o \leq D$ follows $D_o^{m_1} \leq D^{m_1} \leq D^{n_1}$ in contradiction to 2.1(5).
$\overline{\mathfrak{B}}(D) \neq \emptyset$ implies $D \in \overline{\mathfrak{B}}(D)$ in the same way.
2. If $(D^n)_{n \in N}$ is isotonic $(D \leq D^2 \leq \ldots)$ then $\underline{\mathfrak{B}}(D) \neq \emptyset$ and $\overline{\mathfrak{B}}(D) = \emptyset$. In this case (5) may be replaced by

$$(m,n \in N \text{ and } D_o^m \leq D^n) \Rightarrow m \neq n+1.$$

Every Bernoulli system belongs to this case. This is easily shown by the well known factor properties [9] and the entropy relation $h(D^n) = n \cdot h(D)$.
3. If $(D^n)_{n \in N}$ is antitonic $(D \geq D^2 \geq \ldots)$ then $\overline{\mathfrak{B}}(D) \neq \emptyset$ and $\underline{\mathfrak{B}}(D) = \emptyset$. Then (6) may be replaced by

$$(m,n \in N \text{ and } D_o^m \leq D^n) \Rightarrow m \neq n-1. \tag{7}$$

Every ergodic dynamical system D with discrete spectrum belongs to this case. The proper value group H(D) of the corresponding unitary operator is a complete isomorphy invariant and to check (7) one has to use only the fact $H(D^n) = \{h^n \mid h \in H(D)\}$. This and the preceding remark show that the notions of bordering factors are meaningful especially for those D whose sequences $(D^n)_{n \in N}$ are monotonic.

3. Bordering factors in two special classes of dynamical systems

Now we continue the examination via bordering factors in the following two cases: (totally) ergodic systems with discrete spectrum and (especially) Bernoulli systems.

3.1. Theorem. Let D be an ergodic dynamical system with discrete spectrum and with a proper value group $H(D)=\{c_1,\dots,c_k\}$, generated by a finite set of rationally independent $c_1,\dots,c_k$ [1]). Then

(1) every factor $D_o \leq D$ has got a representation

$H(D_o) = \{c_1^{n_1},\dots,c_k^{n_k}\}$ $(n_1,\dots,n_k \in Z_+)$,

(2) $D_o \in \mathfrak{B}(D)$ holds iff the highest common divisor (=h.c.d.) of

$$\{n_i \mid i=1,\dots,k; n_i \neq 0\} \tag{8}$$

equals 1.

Proof. (1): The subgroup property $H(D_o) \subset H(D)$ shows $H(D_o)$ generated by a collection of powers of $c_1,\dots,c_k$ and these may be supposed to be nonnegative.

(2): If the h.c.d. in question equals $d>1$ and if $D_o \leq D$, $H(D_o) = \{c_1^{n_1},\dots,c_k^{n_k}\}$, then $H(D_o) \subset \{c_1^d,\dots,c_k^d\} = H(D^d)$ would hold. The last relation and $d>1$ imply $D_o \notin \mathfrak{B}(D)$.

To prove the converse direction let $D_o^m \leq D^n$ and $m<n$. If the h.c.d. (m,n) of m and n equals $d>1$, then $m=m'd$, $n=n'd$, $H(D_o^{m'd}) \subset H(D^{n'd})$ and therefore $H(D_o^{m'}) \subset H(D^{n'})$. The last implication is equivalent to $D_o^{m'} \leq D^{n'}$. For that reason we may assume altogether

$$D_o^m \leq D^n, \quad 1 \leq m < n, \quad (m,n) = 1. \tag{9}$$

The first of these relations yields

$$H(D_o^m) \subset H(D^n). \tag{10}$$

To prove the contrary of the h.c.d. condition in (8) check the generating elements of $H(D_o)$. Using the rational independence assumption on $c_1,\dots,c_k$ (10) effects $c_i^{n_i m} = c_i^{l_i n}$ and (9) yields $n \mid n_i$ for all n_i belonging to (8).

3.2. Corollary. Every totally ergodic dynamical system D with discrete

---

[1]) This implies total ergodicity of D (D^n ergodic for all $n \in N$).

spectrum and $H(D)=\{c\}$ has no D-bordering factor from above, which is different from D.

3.3. Remark. If D is an ergodic but not totally ergodic dynamical system with discrete spectrum and proper value group $H(D)=\{c\}$, then $\mathfrak{F}(D)=\mathfrak{B}(D)=\emptyset$. This follows from the existence of a natural $k>0$ with $c^k=1$. Then D^{k+1} is ergodic, with discrete spectrum and with the proper value group $H(D^{k+1})=\{c\}$ and is isomorphic to D. Therefore $D\notin\mathfrak{F}(D)$ because $D\leq D^{k+1}$ and $D\notin\mathfrak{B}(D)$ because $D^{k+1}\leq D$.

3.4. Remark. For any dynamical system D with positive entropy there exists a Bernoulli system D_o with $h(D_o)=h(D)$, $D_o\leq D$ [11]. This D_o is D-bordering from below otherwise

$$D_o^{m_1}\leq D^{n_1} \text{ and } m_1>n_1 \tag{11}$$

would hold for some $m_1,n_1\in N$. In this case we would estimate

$$m_1 h(D_o) = h(D_o^{m_1})\leq h(D^{n_1}) \Rightarrow m_1\leq n_1$$

in contradiction to (11).

Therefore every Kolmogorov system D has one D-bordering factor from below at least. The following theorem deals with uniqueness of this bordering factor (mod conj).

3.5. Theorem. A dynamical system D with positive entropy is Bernoulli iff D itself is the only D-bordering factor from below (mod conj).

Proof. 1. Let D be Bernoulli and $D_o\leq D$ and D-bordering. Then the corresponding entropies satisfy $h(D_o)\leq h(D)$ and D_o is Bernoulli, too [9]. In case that $h(D_o)<h(D)$ an integer n would exist with $h(D_o^n)=n\cdot h(D_o)\leq$ $\leq(n-1)\cdot h(D)=h(D^{n-1})$ and therefore $D_o^n\leq D^{n-1}$ contradictory to (5). Therefore $h(D_o)=h(D)$ implies conjugacy of D,D_o.

2. Let conversely

$$(D_o\leq D,\ D_o \text{ D-bordering from below}) \Rightarrow D_o \text{ conj } D \tag{12}$$

hold. Then (see remark 3.4) the Bernoulli system D_o with entropy $h(D_o)=$ $=h(D)$ satisfies the condition of (12). Consequently D is a Bernoulli system mod conj.

3.6. Corollary. A Kolmogorov system D is Bernoulli if and only if D itself is the only D-bordering factor from below.

4. Narrowly bordering factors

We refine the notion of D-bordering as indicated in the introduction. In terms of motivating functions bordering includes contact of the graphs of f and g in possibly more than one point but in the refined definition touch is allowed only in one point. Again it is essential to avoid the notion of point in the dynamical systems case.

4.1. Definition. For any two dynamical systems D_1, D_2 let

$$\mathcal{F}(D_1,D_2) = \{ D \mid D \leq D_i \ (i=1,2)\}$$

the class of lower bounds of D_1, D_2 and

$$\mathcal{F}_*(D_1,D_2) = \{D \in \mathcal{F}(D_1,D_2) \mid (D' \in \mathcal{F}(D_1,D_2) \text{ with } D \leq D') \Rightarrow D' \leq D\}$$

the class of maximal lower bounds of D_1, D_2.

4.2. Remarks. 1. This definition is well in the following sense. The preorder defined above implies an order $\dot{\leq}$ in the class of (weakly isomorphic) equivalence classes

$$[D] = \{D' \mid D' \leq D \text{ and } D \leq D'\}.$$

Then $D \in \mathcal{F}_*(D_1,D_2)$ means maximality of $[D]$ in the usual sense in

$$\mathcal{F}([D_1],[D_2]) = \{[D] \mid [D] \dot{\leq} [D_i] \ (i=1,2)\}.$$

2. The property $\mathcal{F}_*(D_1,D_2) \neq \emptyset$ is known or easy to prove for D_1, D_2 belonging to some special classes of dynamical systems with well known structure. The class $\mathcal{B}$ of Bernoulli systems is an illustrative example. For $D \in \mathcal{B}$ every factor is Bernoulli again [9], and the entropy h(D) is a complete isomorphy invariant for these systems. One obtains

$$\mathcal{F}(D_1,D_2) = \{ D \in \mathcal{B} \mid h(D) \leq \min(h(D_1),h(D_2))\},$$

$$\mathcal{F}_*(D_1,D_2) = \{ D \in \mathcal{F}(D_1,D_2) \mid h(D) = \min(h(D_1),h(D_2))\}.$$

Consequently, for two Bernoulli systems the maximal lower bound is uniquely determined (mod isom).

For D_1, D_2 from the class $\mathcal{K}$ of all ergodic dynamical systems with discrete spectrum we have

$$\mathcal{F}(D_1,D_2) = \{D \in \mathcal{K} \mid H(D) \subset H(D_1) \cap H(D_2)\},$$

$$\mathcal{T}_*(D_1,D_2) = \{D \in \mathcal{T}(D_1,D_2) \mid H(D)=H(D_1) \cap H(D_2)\},$$

where H(D) is the group of proper values of D. Thus in this class of dynamical systems one can describe the classes $\mathcal{T}(D_1,D_2)$ and $\mathcal{T}_*(D_1,D_2)$ making use of the complete isomorphy invariant [6] of proper values and again $\mathcal{T}_*(D_1,D_2)$ contains only one element (mod isom).

Replacing $\mathcal{K}$ by the class $\mathcal{K}^*$ of totally ergodic systems with quasidiscrete spectrum any $\mathcal{T}_*(D_1,D_2)$ with $D_1,D_2 \in \mathcal{K}^*$ contains only one element (mod isom), too [4].

3. In the following we use $\mathcal{T}_*(D_1,D_2)$ only in a rather implicit sense.

4.3. Definition. A D-bordering factor D_o is called narrowly D-bordering from below if

$$D_1,D_2 \in \mathcal{T}(D_o) \cap \underline{\mathfrak{B}}(D) \Rightarrow \emptyset \neq \mathcal{T}_*(D_1,D_2) \subset \underline{\mathfrak{B}}(D), \tag{13}$$

resp. narrowly D-bordering from above if

$$D_1,D_2 \in \mathcal{T}(D_o) \cap \bar{\mathfrak{B}}(D) \Rightarrow \emptyset \neq \mathcal{T}_*(D_1,D_2) \subset \bar{\mathfrak{B}}(D). \tag{14}$$

Thereby we use the notation $\mathcal{T}(D)=\{D' \mid D' \leq D\}$ for the class of all factors of D. The classes of narrowly D-bordering factors from below (from above) are denoted by $\underline{\mathfrak{B}}_N(D)$ ($\bar{\mathfrak{B}}_N(D)$).

4.4. Remarks. 1. For every system D with positive entropy h(D) the Bernoulli system D_o with $h(D_o)=h(D)$ is narrowly D-bordering from below (see 3.4 and 3.5) and for every ergodic dynamical system D with discrete spectrum determined by the proper value group $H(D)=\{c_1,\dots,c_k\}$, where $c_1,\dots,c_k \in K=\{z \in C \mid |z|=1\}$ are rationally independent, all ergodic systems D_i with discrete spectrum and proper value group $H(D_i)=\{c_i\}$ and only these are narrowly D-bordering from above (i=1,...,k),(see 3.1.) .

2. There exist dynamical systems D without narrowly D-bordering factors because $\underline{\mathfrak{B}}(D) = \bar{\mathfrak{B}}(D) = \emptyset$ is possible (see 3.3).

3. It seems to be more complicated whether there exist dynamical systems D with the property $\underline{\mathfrak{B}}(D)\neq\emptyset$ and $\underline{\mathfrak{B}}_N(D)=\emptyset$ (resp. $\bar{\mathfrak{B}}(D)\neq\emptyset$ and $\bar{\mathfrak{B}}_N(D)=\emptyset$).

With respect to our motivation the following property should be and is valid:

4.5. Theorem. Any D-bordering factor D_{oo} of a narrowly D-bordering factor D_o of D is narrowly D-bordering:

$$\mathfrak{B}(D) \cap \mathcal{F}(\mathfrak{B}_N(D)) \subset \mathfrak{B}_N(D).$$

Proof. (13) and (below case) $D_{oo} \in \mathfrak{B}(D)$ are assumed. We have to show

$$D_{1o}, D_{2o} \in \mathcal{F}(D_{oo}) \cap \mathfrak{B}(D) \Rightarrow \mathcal{F}_*(D_{1o}, D_{2o}) \subset \mathfrak{B}(D).$$

Since $\mathcal{F}(D_{oo}) \subset \mathcal{F}(D_o)$ this follows from (13).

4.6. Theorem. If $D_o \in \mathfrak{B}(D)$ and if D_o has no proper D-bordering factors (mod conj), so D_o is narrowly D-bordering.

Proof. The assumptions $D_1, D_2 \in \mathcal{F}(D_o) \cap \mathfrak{B}(D)$ imply $D_1 = D_2 = D_o$ (mod conj) and therefore $\mathcal{F}_*(D_1, D_2) = \{D_o\} \subset \mathfrak{B}(D)$.

4.7. Remark. 4.4.1 shows that there exist dynamical systems D with more than one narrowly D-bordering factor. The narrowly D-bordering factors of the ergodic dynamical system D with discrete spectrum from 4.4.2 have a property of extreme difference: any two systems of $\bar{\mathfrak{B}}_N(D)$ have no nontrivial common factor (see [3]).

4.8. Theorem. A dynamical system with positive entropy is Bernoulli iff D is the only narrowly D-bordering factor from below.

Proof. 1. If D is Bernoulli then by 3.5 D is the only D-bordering factor from below of D (mod conj). Therefore by 4.6 D is narrowly D-bordering from below. The uniqueness of the narrowly D-bordering factor follows from the uniqueness of D-bordering.

2. Let D be the only narrowly D-bordering factor from below. The existence of another (not weakly isomorphic to D) D-bordering factor D_o of D would imply a contradiction because there is theorem 4.5.

4.9. Remark. There exist examples of dynamical systems D with the following properties:

(1) $\emptyset \neq \bar{\mathfrak{B}}_N(D)$ $(\subset \bar{\mathfrak{B}}(D))$,

(2) there exists an antitone sequence $(D_i)_{i \in N} \subset \bar{\mathfrak{B}}(D)$ such that

$$\bigcap_{i \in N} \mathcal{F}(D_i) \cap \bar{\mathfrak{B}}_N(D) = \emptyset.$$

It is sufficient, to choose D as totally ergodic dynamical system with discrete spectrum defined by the proper value group

$$H(D) = \{c_1, c_2, \ldots\}$$

where $c_i \in K$ ($i \in N$) and every finite subsystem of the c_i is rationally independent. Then the sequence $(D_i)_{i \in N}$ of the totally ergodic systems with discrete spectrum and proper value groups

$$H(D_i) = \{c_i, c_{i+1}, \ldots\}$$

has the obvious properties

$$D_{i+1} \leq D_i,\ D_i \in \bar{\mathfrak{B}}(D) \quad (i \in N).$$

But no system D_i is an element of $\bar{\mathfrak{B}}_N(D)$. Namely, the relatively prime factors $D_i^{(1)}, D_i^{(2)}$ of D_j, which are ergodic, with discrete spectrum and defined by

$$H(D_i^{(1)}) = \{c_i\},\ H(D_i^{(2)}) = \{c_{i+1}\},$$

are D-bordering from above, therefore $\bigcup_{i \in N} \{D_i\} \cap \bar{\mathfrak{B}}_N(D) = \emptyset$. Now we suppose that there exists a system D_o in $\bigcap_{i \in N} \mathcal{F}(D_i) \cap \bar{\mathfrak{B}}_N(D)$. In this case D_o would be an element in the class of all ergodic dynamical systems with discrete spectrum (because factors are in the same class, see [6]), $H(D_o) \subset H(D_j)$ ($j \in N$) and $H(D_o) \neq \{1\}$, since D_o is assumed as a D-bordering factor of the nontrivial system D. If $c \in H(D_o)$ with $c \neq 1$, then by definition of D there exist finite many integers $i_1 < \ldots < i_k$ such that $c = c_{i_1}^{m_1} c_{i_2}^{m_2} \cdots c_{i_k}^{m_k}$. The generating elements of the $D_{i_k+1} \in (D_i)_{i \in N}$ are rationally independent from c. Therefore we have $H(D_o) \setminus H(D_{i_k+1}) \neq \emptyset$, i.e. $H(D_o) \not\subset H(D_{i_k+1})$. But this is in contradiction to $D_o \leq D_{i_k+1}$.

Now we construct a sort of equivalence classes in $\underline{\mathfrak{B}}_N(D)$ (resp. $\bar{\mathfrak{B}}_N(D)$).

__4.10. Definition.__ Two systems $D_1, D_2 \in \underline{\mathfrak{B}}_N(D)$ are called equivalent ($D_1 \asymp D_2$) if there exists a D' with $D' \leq D_1, D' \leq D_2$ and $D' \in \underline{\mathfrak{B}}(D)$.

In our terminology this means $\mathcal{F}(D_1, D_2) \cap \underline{\mathfrak{B}}(D) \neq \emptyset$. Of course, this is an equivalence relation. Only

$$D_1 \asymp D_2,\ D_2 \asymp D_3 \Rightarrow D_1 \asymp D_3$$

is worth to be proved. Let $D_{12} \in \mathcal{F}(D_1, D_2) \cap \underline{\mathfrak{B}}(D)$, $D_{23} \in \mathcal{F}(D_1, D_2) \cap \underline{\mathfrak{B}}(D)$ and $D_{13} \in \mathcal{F}_u(D_{12}, D_{23})$. Then $D_{13} \leq D_{12} \leq D_1$, $D_{13} \leq D_{23} \leq D_3$ imply

$$D_{13} \in \mathcal{F}(D_1, D_3). \tag{15}$$

D_2 is narrowly D-bordering. Therefore we have with (13)

$$D', D'' \in \mathcal{F}(D_2) \cap \underline{\mathfrak{B}}(D) \Rightarrow \mathcal{F}_*(D', D'') \subset \underline{\mathfrak{B}}(D).$$

Instead of D', D'' the dynamical systems D_{12}, D_{23} are available and $\mathcal{F}_*(D_{12}, D_{23}) \subset \underline{\mathfrak{B}}(D)$, therefore $D_{13} \in \underline{\mathfrak{B}}(D)$ follows. Together with (15) this yields $D_1 \asymp D_3$.

4.11. Remark. It is obvious that for weakly isomorphic systems $D_1, D_2 \in \underline{\mathfrak{B}}_N(D)$ (resp. $\in \overline{\mathfrak{B}}_N(D)$) holds $D_1 \asymp D_2$.

4.12. Example. If D is an ergodic dynamical system with discrete spectrum and the proper value group $H(D) = \{c_1, \ldots, c_k\}$, where $c_i \in K$ and rationally independent $(i=1,\ldots,k)$, then it holds

$$\overline{\mathfrak{B}}_N(D) = \bigcup_{i=1}^{k} \langle D_i \rangle.$$

where $\langle D_i \rangle$ is the equivalence class with respect to the $\asymp$-relation in $\overline{\mathfrak{B}}_N(D)$, generated by the ergodic dynamical system D_i with discrete spectrum and $H(D_i) = \{c_i\}, (i=1,\ldots,k)$.

REFERENCES

[1] Clark, J.: A Kolmogorov shift with no roots. Ph. D. Thesis, Stanford Univ., Calif. 1972.

[2] Chacon, R.V.: A geometric construction of measure preserving transformations. Proc. Fifth Berkeley Sympos. Math. Stat. and Probab., Berkeley and Los Angeles (1965), vol. 2, 335-360.

[3] Furstenberg, H.: Disjointness in ergodic theory, minimal sets and a problem in Diophantine approximation. Math. Systems Theory 1 (1967), 1-49.

[4] Häsler, K.: Obere und untere Schranken vollergodischer dynamischer Systeme mit quasidiskretem Spektrum. To appear in: Zeitschrift für Analysis und ihre Anwendungen 1(1), 1982.

[5] Hahn, F. and W. Parry: Some characteristic properties of dynamical

systems with quasidiscrete spectrá. Math. Systems Theory 2 (1968) 179-190.

[6] Jacobs,K.: Lecture Notes in Ergodic Theory. Part I,II. Aarhus 1962/63.

[7] Michel,H.: Support as an invariant for dynamical systems. Math. Slovaca 30, 1980, no. 1, 19-30.

[8] Michel,H.: Vollergodische Automorphismen mit quasidiskretem Spektrum und Wurzeln jeder rationalen Ordnung. Beiträge zur Analysis und Angewandten Mathematik, Wiss. Zeitschr. 1968/9 (M1) Univ. Halle, 123-128.

[9] Ornstein,D.: Factors of Bernoulli shifts are Bernoulli shifts. Adv. Math. 5 (1971), 349-364.

[10] Ornstein,D.: A K-automorphism with no square root and Pinsker's conjecture. Adv. Math. 10 (1973), 89-102.

[11] Sinai,Ja.G.: On a weak isomorphism of transformations with invariant measure (Russian). Math. Sb. (N.S.) 63 (105) (1964), 23-42.

ENTROPY OF SIMPLE BROKEN LINEAR FUNCTIONS

Gilbert Helmberg [1)]

1. Introduction and main results

We are concerned with transformations $f = f(s,\bar{s})$ of the unit interval $[0,1]$ given by

$$f(x) = f(s,\bar{s};x) = \begin{cases} sx & \text{for } 0 \le x \le \frac{\bar{s}}{s+\bar{s}} \\ \bar{s}(1-x) & \text{for } \frac{\bar{s}}{s+\bar{s}} \le x \le 1 \end{cases} \tag{1}$$

$(0<s<\infty,\ 0<\bar{s}<\infty,\ (s-1)(\bar{s}-1)\le 1)$ (figure 1). The topological entropy of such a function is given by

$$h(f) = h(s,\bar{s}) = \lim_{n\to\infty} \frac{1}{n} \log c(f^{(n)}) = \lim_{n\to\infty} \frac{1}{n} \log \operatorname{var}(f^{(n)}) \tag{2}$$

(Misiurewicz/Szlenk [1977,1980], Roider [1978]). Here $f^{(n)}$ signifies the n-th iterate of f ($f^{(1)} = f$, $f^{(n+1)} = f \circ f^{(n)}$), $c(f^{(n)})$ is the number of maximal monotony-intervals of $f^{(n)}$ (figure 3) and $\operatorname{var}(f^{(n)})$ means the total variation of $f^{(n)}$.

The peak point P of the graph of $f(s,\bar{s})$ is given either by $P(s,\bar{s})$ or by its Cartesian coordinates (p,q) $(s = q/p,\ \bar{s} = q/(1-p))$. We are interested in curves of constant entropy h

$C(h) = \{P(s,\bar{s}) : h(s,\bar{s}) = h\}$.

A simple example is given by the line $y = 1$ $(\bar{s} = s/(s-1))$. For every corresponding function $f(s,s/(s-1))$ $(1<s<\infty)$ one readily obtains $c(f^{(n)}) = 2^n$ and therefore $h(s,s/(s-1)) = \log 2$.

On the other hand, for $p = 1/2$ $(s=\bar{s})$ one again readily obtains $\operatorname{var}(f^{(n)}) = s^n$ and therefore $h(s,\bar{s}) = \log s$ $(1\le s\le 2)$. Thus, every value $0 \le h \le \log 2$ certainly occurs as entropy for functions of type (1). Finally, for $0 < s \le 1$ or $0 < \bar{s} \le 1$ the function f is seen to act periodically on its non-wandering set, so by general theorems (Denker/Grillenberger/Sigmund [1976]) one has $h(s,\bar{s}) = 0$ in this case.

As a consequence, curves of constant non-zero entropy may turn up only in the triangle bounded by the line $y = 1$ and both diagonals of the unit square. However, despite of this seemingly symmetric set-up, the configuration formed by these curves turns out to be rather asymmetric. In order to describe it consider first the region D bounded by the line $y = 1$, the diagonal $y = x$, and by the parabola $y^2 = 1-x$

$$D = \{P(s,\bar{s}) : \frac{1+\sqrt{1+4s^2}}{2s} \le \bar{s} \le \frac{s}{s-1},\ 1<s<\infty\}.$$

Theorem 1 For $P(s_0,\bar{s}_0) \in D$ one has $h(s_0,\bar{s}_0) \in [\log\sqrt{2}, \log 2]$. $C(h(s_0,\bar{s}_0))$ is a continous curve contained in D and intersecting every line

---

1) Institut für Mathematik und Geometrie, Technische Fakultät, Technikerstraße 13, A-6020 Innsbruck, Austria

$y = sx$ $(1<s<\infty)$ in precisely one point $P(s,\bar{s}(s))$, for which moreover

$$\lim_{s\to\infty} P(s,\bar{s}(s)) = (0,1) \text{ and}$$

$$\lim_{s\searrow 1} P(s,\bar{s}(s)) \text{ exists.}$$

Secondly consider the injective transformation of the unit square $T:(x,y) \to (1-x,\sqrt{xy})$ also given by

$$TP(s,\bar{s}) = P(\bar{s}/\sqrt{s},\sqrt{s})$$

(figure 6). Its inverse is given on the image set by

$$T^{-1}P(s,\bar{s}) = P(\bar{s}^2,s\bar{s}).$$

Let

$$c_o = \{(x,y):y=1,\ 0<x<1\} = \{P(s,\tfrac{s}{s-1}):\ 1<s<\infty\} \qquad (3)$$

and observe that T maps the segment

$$c_o \cap\{P(s,\bar{s}):1<s<\bar{s}^2\} = \{P(s,\tfrac{s}{s-1}):\ 1<s<\tfrac{3+\sqrt{5}}{2}\}$$

onto the parabolic arc

$$c_1 = \{P(\tfrac{\bar{s}}{\bar{s}^2-1},\bar{s})\ :\ \tfrac{1+\sqrt{5}}{2}<\bar{s}<\infty\} \qquad (4)$$

(part of the parabola $y^2 = 1-x$) forming the "lower left boundary" of D. Inductively one may form the arcs

$$c_{i+1} = \{TP(s,\bar{s}):\ P(s,\bar{s}) \in c_i,\ 1<s<\bar{s}^2\}$$

which join both diagonals $s = 1$ and $\bar{s} = 1$. Denote by D_i the region bounded by these diagonals (exclusively) and the arcs c_i (inclusively) and c_{i+1} (exclusively) (figure 7; note that $D_o = D\setminus c_1$).

<u>Theorem 2</u>. For $2^{-(i+1)}\log 2 < h \leq 2^{-i}\log 2$ $(1\leq i<\infty)$ one has

$$C(h) = \{TP(s,\bar{s}):\ P(s,\bar{s}) \in C(2h),\ 1<s<\bar{s}^2\} \subset D_i$$

In what follows we clarify (by a rather geometrically oriented approach) the connection between the entropy $h(s,\bar{s})$ and the orbit configuration of $P(s,\bar{s})$ which forms the basis for these results (theorem 3). Applied to periodic orbit configurations this connection allows the construction of algebraic curves of constant entropy. Finally we explain the idea of proof of theorem 1 and the reason for the recursive character of theorem 2. Detailed proofs may be found in a more elaborate version of this paper to be published in Monatshefte für Mathematik.

2. The orbit configuration of $P(s,\bar{s})$

The clue for obtaining points with the same corresponding entropy lies in a result essentially due to Milnor/Thurston [1977], Derrida/Gervais/Pomeau [1978], Jonker/Rand [1980] and based on the first formula in (2). Given $P(s,\bar{s}) = (p,q)\in D$ consider the sequence

$$\varepsilon = \varepsilon(s,\bar{s}) = \{\varepsilon_k(p) = \operatorname{sign}(p-f^{(k)}(p))\}_{k=0}^{\infty} \qquad (5)$$

(sequences of this type have already been considered by Metropolis/Stein/Stein [1973]). Evidently one has $\varepsilon_o = 0$, $\varepsilon_1 = -1$, $\varepsilon_2 = +1$.

<u>Theorem 3</u>. If $\varepsilon(s,\bar{s}) = \varepsilon(s_o,\bar{s}_o)$, then $h(s,\bar{s}) = h(s_o,\bar{s}_o)$.

<u>Proof</u>. The following statements on the graph Γ_n of $f^{(n)}$ may be verified inductively but become rather evident if f and $f^{(k)}$ are composed to $f^{(k)}$of as e.g. in figure 3.

a) Γ_n is connected and piecewise linear; its corners lie on the levels $y_k = f^{(k)}(p)$ $(1 \leq k \leq n)$.

b) As a saw-tooth configuration with these corner levels Γ_n appears to be symmetric with the symmetry axis passing through the point $P^{(n)} = (p, f^{(n)}(p))$.

c) To the left of this axes Γ_n contains the corner points $P_k^{(n)} = (p.(p/q)^{n-k}, f^{(k)}(p))$ for $1 \leq k \leq n$. Between 0 and $P_1^{(n)}$ also $P_o^{(n)} = (p(p/q)^n, y_o = p)$ lies on Γ_n.

d) The part of Γ_n between $P_{k-1}^{(n)}$ and $P_k^{(n)}$ $(1 \leq k \leq n)$ is also obtained by composing $f^{(n-1)}$ and f to fof$^{(n-1)}$ in the interval $[p(p/q)^{n-k+1}, p(p/q)^{n-k}]$. Every segment of Γ_{n-1} between the levels, say, y_i and y_j, and meeting the line $y = y_o$ in its interior, gives rise to two segments of Γ_n between the levels y_{i+1} and y_1 on the one hand, respectively y_1 and y_{j+1} on the other hand. Every other segment of Γ_{n-1} gives rise to just one segment of Γ_n between the levels y_{i+1} and y_{j+1}.

e) This implies that the number $c(f^{(n)})$ of segments of Γ_n is already determined by the position of y_k $(1 \leq k \leq n-1)$ with respect to $y_o = p$ $(y_k > y_o, y_k = y_o, y_k < y_o)$. Using (5) and (2) one obtains theorem 3.

3. <u>Algebraic curves of constant entropy</u>

Let

$$v_k = \frac{f^{(k)}(p)-p}{q-p} \qquad (0 \leq k \leq \infty)$$

(including $v_o = 0$) A routine computation reveals

$$v_{k+1} = \begin{cases} 1 + sv_k & \text{if } \varepsilon_k(p) = +1 \\ 1 - \bar{s}v_k & \text{if } \varepsilon_k(p) = -1 \end{cases}$$

Suppose $\varepsilon_i \neq 0$ for $1 \leq i \leq k$ and let $\theta_k = \prod_{i=1}^{k} \varepsilon_i$, $\theta_o = 1$. Then

$$\theta_k v_{k+1} = \sum_{i=0}^{k} \theta_{k-i}\, s^{a(i,k)} \bar{s}^{i-a(i,k)}$$

where $a(0,k) = 0$ and (for $1 \leq i \leq k$) $a(i,k)$ is the number of positive signs among the ε_j's with $k-i+1 \leq j \leq k$.

The case $\varepsilon_n = 0$ (in which the sequence ε becomes periodic with period n) is therefore characterized by

$$\sum_{i=0}^{n-1} \theta_{n-1-i}\, s^{a(i,n-1)} \bar{s}^{i-a(i,n-1)} = 0, \qquad (6)$$

an algebraic equation of degree n-1 in $s,\bar{s}$ (or p,q if expressed in Cartesian coordinates). All points $P(s,\bar{s})$ admitting the same sequence ε (and therefore the same entropy by theorem 3) have to satisfy this equation.

In particular we obtain points of equal corresponding entropy if we determine s as a function of $\bar{s}$ (or vice versa) in such a way that $\varepsilon_1 = -1$ and $\varepsilon_2 = +1$ are followed by a given number of k ($\leq\infty$) equal signs (and for $k<\infty$ then by a zero). Elementary geometric considerations (figure 2) show that this requirement is equivalent to the following equations to be satisfied by $v_2 = 1-\bar{s}$ resp. $v_3 = 1+s-s\bar{s}$. The corresponding entropy is then given by log s_o where $P(s_o,s_o)$ (with $p_o = 1/2$) satisfies the corresponding equation in $s,\bar{s}$.

1) $\varepsilon_i = +1$ for all $i \geq 2$ $(v_2 = v_{-\infty} = \frac{1}{1-s})$:

$$\bar{s} = \frac{s}{s-1}, \quad h(s,\bar{s}) = \log 2$$

$P(s,\bar{s})$ lies on c_o (3).

2) $\varepsilon_i = +1$ for $2 \leq i \leq k+1$, $\varepsilon_{k+2} = 0$ $(v_2 = v_{-k} = \frac{1-s^{-k}}{1-s})$:

$$\bar{s} = \frac{s-s^{-k}}{s-1}$$

The entropy $h(s,\bar{s})$ is the logarithm of the root of the equation

$$2 - s = \frac{1}{s^{k+1}}$$

(figure 4). As $k \to \infty$ the entropy increases towards log 2.

A special case is obtained for $k = 1$, namely an ellipse with the equation

$$p^2+pq+q^2-p-q = 0$$

and with corresponding entropy $\log \frac{1+\sqrt{5}}{2}$ (figure 5).

3) $\varepsilon_i = -1$ for $3 \leq i \leq k+2$, $\varepsilon_{k+3} = 0$ $(v_3 = \bar{v}_{-k} = \frac{1-(-\bar{s})^{-k}}{1+\bar{s}})$:

$$s = \frac{\bar{s}+(-\bar{s})^{-k}}{\bar{s}^2-1}.$$

The corresponding entropy is the logarithm of the root of the equation

$$(-1)^k(s^2-2) = \frac{1}{s^{k+1}} \qquad \text{(figure 4).}$$

a) k even: as $k \to \infty$ the entropy decreases from $\log \frac{1+\sqrt{5}}{2}$ to $\log \sqrt{2}$.

b) k odd: as $k \to \infty$ the entropy increases to $\log \sqrt{2}$.

4) $\varepsilon_i = -1$ for all $i \geq 3$ $(v_3 = \bar{v}_{-\infty} = \frac{1}{1+\bar{s}})$:

$$s = \frac{\bar{s}}{\bar{s}^2-1}, \quad h(s,\bar{s}) = \log \sqrt{2}.$$

This equation describes the parabola $q^2 = 1-p$ (4), representing the limiting case between 3a) und 3b).

In the cases 1), 2), 3a), and 4) one has

$$\lim_{s\to\infty} \bar{s} = 1, \quad \lim_{\bar{s}\searrow 1} s = \infty,$$

in case 3b), however

$$\lim_{\bar{s}\searrow 1} s = \frac{k+1}{2}.$$

Furthermore one has

$$\lim_{s \searrow 1} \bar{s} = \begin{cases} \infty & \text{in case 1)} \\ k+1 & \text{in case 2)} \\ \bar{s}_k & \text{in case 3)} \\ \frac{1+\sqrt{5}}{2} & \text{in case 4)} \end{cases}$$

where $\bar{s}_k$ is the root (between 1 and 2) of the equation

$$(-1)^k(\bar{s}^2-\bar{s}-1) = \frac{1}{\bar{s}^k} \qquad \text{(figure 4).}$$

Finally, a routine calculation shows that the curves of constant entropy in case 1),2) and 3a) are tangential to c_o (y=1) in the point (0,1).

4. Sketches of proofs of theorem 1 and theorem 2

The proof of theorem 1 (at least the one I am able to give) is technically too involved to be reproduced here. The idea consists in the following steps.

1) For given non-zero $\varepsilon_1,\ldots,\varepsilon_{n-1}$ show that $\frac{\partial v_n}{\partial \bar{s}}$ is bounded away from O in D. This implies that (6) locally defines $\bar{s}$ as a continous (even differentiable) function of s in a neighbourhood of any point P ∈ D associated with the given sequence $\{\varepsilon_o,\varepsilon_1,\ldots,\varepsilon_{n-1}\}$.

2) Show that $\varepsilon_1,\ldots,\varepsilon_{n-1}$ remain non-zero along any continuous extension of the graph of such a function unless it ends in (0,1) or on the diagonal s = 1. This implies that (6) globally defines a curve of constant entropy enjoying the properties mentioned in theorem 1.

3) Show that for any non-periodic $\varepsilon=\varepsilon(s_o,\bar{s}_o)$ the set of all points $P(s,s) \in D$ such that $\varepsilon(s,\bar{s})=\varepsilon$ may be uniformly approximated in D by algebraic curves of type (6). It therefore inherits all properties just mentioned.

4) Each of these curves must intersect the line $x = \frac{1}{2}$ for continuity reasons. Since entropy is constant along this curve it can intersect the line $x = \frac{1}{2}$ only in one point which then indicates this entropy.

As to theorem 2, an inspection reveals that $f^{(2)}(s,\bar{s})$ leaves two disjoint intervals invariant which contain its non-wandering set (figure 8). On each of these intervals $f^{(2)}$ induces a transformation of type $f(\bar{s}^2,s\bar{s})$ (this observation has been communicated to me by F. Hofbauer). Since $f^{(2)}$ has entropy $2h(s,\bar{s})$ one has $2h(s,\bar{s}) = h(\bar{s}^2,s\bar{s})$ and $C(h)\subset TC(2h)$ The proof of theorem 2 may then be completed by observing that the part of D_i to the right of the hyperbola $s = \bar{s}^2$ is mapped by T onto D_{i+1}.

As corollaries one obtains continuity of the entropy inside of the triangle bounded by c_o and both diagonals. These diagonals are lines of discontinuity expect for the center of the unit square. Moreover $h(s,\bar{s})$ is strictly monotonic in s and $\bar{s}$ in this triangle.

Literature

Denker, M./Grillenberger,C./Sigmund,K. [1976]
Ergodic theory on compact spaces.
Lecture Notes in Mathematics 527, Springer Verlag, Berlin 1976.

Derrida,B./Gervais,A./Pomeau,Y. [1978],
Iteration of endomorphisms on the real axis and representation of numbers.
Ann.Inst. Henri Poincaré 29 (1978), 305-356.

Jonker,L./Rand,D. [1980]
The periodic orbits and entropy of certain maps of the unit interval.
J. London Math. Soc. (2) 22, (1980), 175-181.

Metropolis,N./Stein,M.L./Stein,P.R. [1973]
On finite limit sets for transformations.
J. of combinatorial theory (A) 15 (1973), 25-44.

Milnor,J./Thurston,W. [1977]
On iterated maps of the interval I,II. Preprint.
Princeton University and the Institute for Advanced Studies, Princeton 1977.

Misiurewicz,M./Szlenk,W. [1977]
Entropy of piecewise monotone mappings.
Astérisque 50 (1977), 299-310.

Misiurewicz,M./Szlenk,W. [1980]
Entropy of piecewise monotone mappings.
Studia math. 67(1980), 45-63.

Roider,B. [1978]
One-dimensional considerations on topological entropy.
Institut f. Mathematik I, Univ. Innsbruck,
Studienjahr 1977/78, Seminarnotiz Nr. 5.

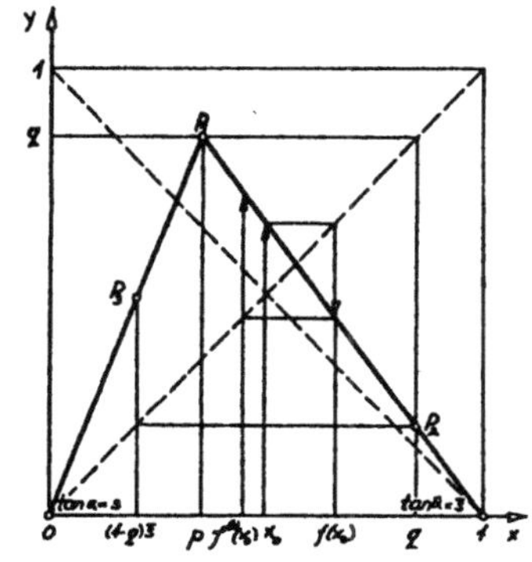

Figure 1

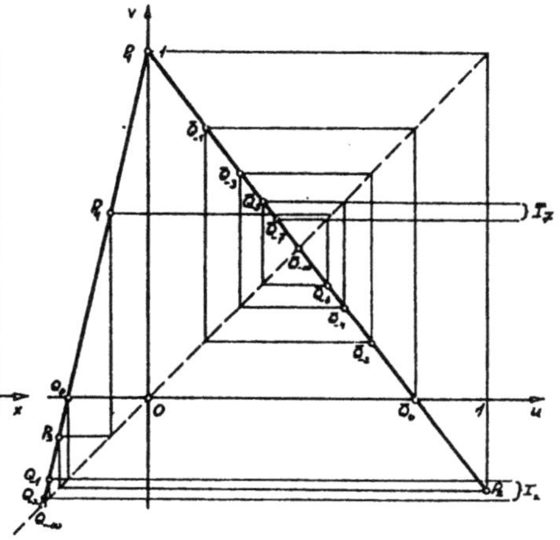

Figure 2

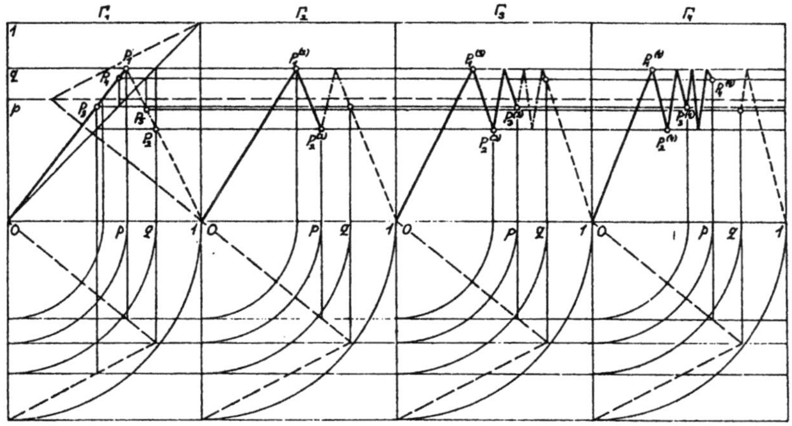

Figure 3

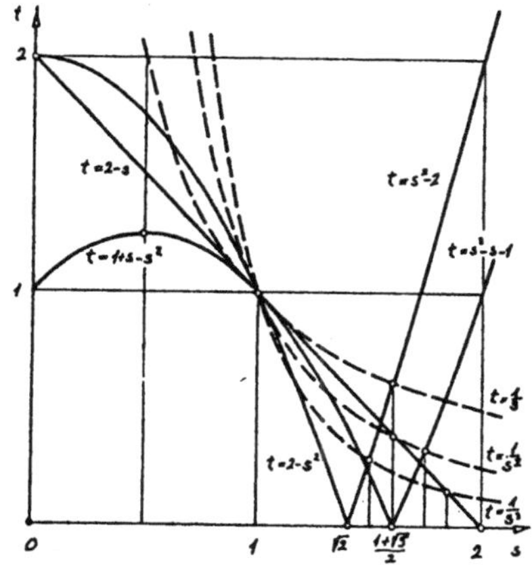

Figure 4

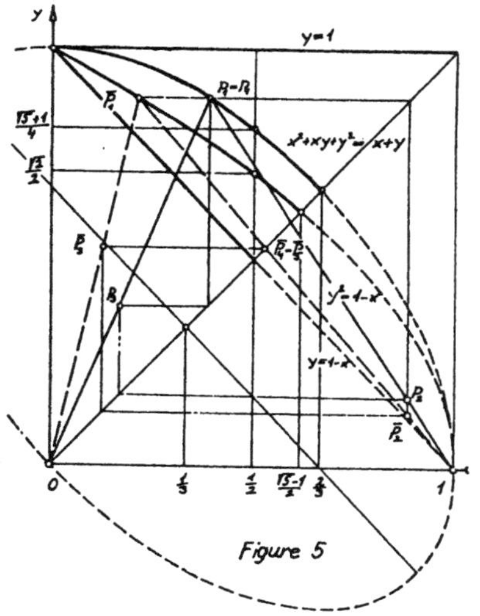

Figure 5

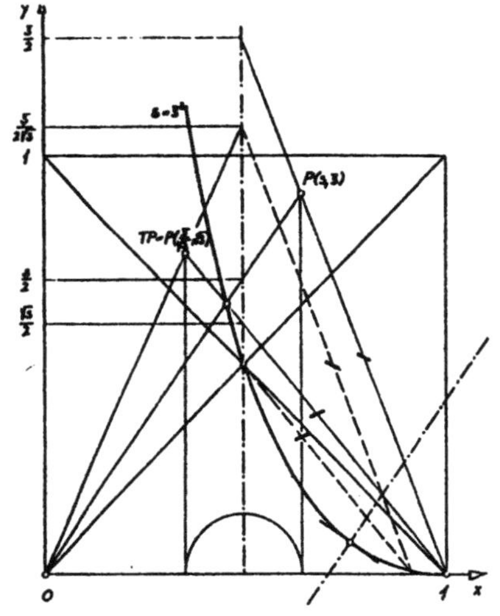

Figure 6

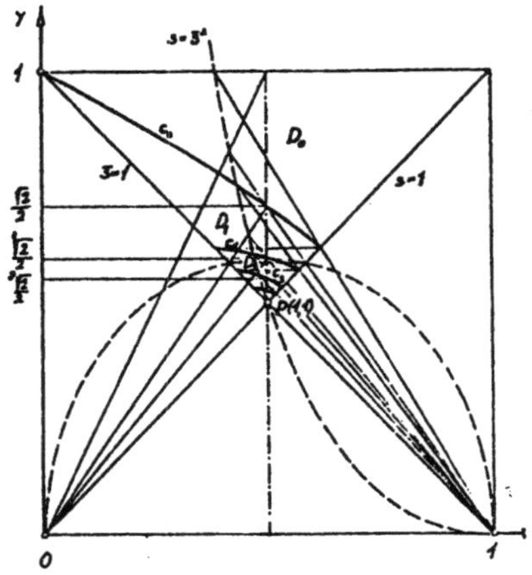

Figure 7

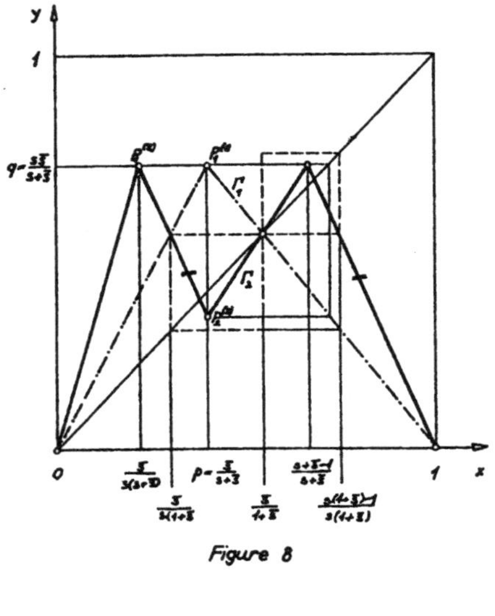

Figure 8

KNEADING INVARIANTS AND MARKOV DIAGRAMS

Franz Hofbauer[1)]

§ 0 . Introduction.

In this paper we want to compare two methods for the investigation of the nonwandering set of piecewise monotonic transformations. The papers [2], [3] and [4] deal with general piecewise monotonic transformations. The main tool used there is an oriented graph which we shall call the Markov diagram of the transformation. In [6] the class of continuous transformations f on [0,1] is studied, which have a unique maximum in $c \in (0,1)$, satisfy $f(0) = f(1) = 0$, and are strictly monotone on either side of c. Using a formal power series, called kneading invariant, stronger results than those of [4] are proved for these transformations f.

In the sequel we shall apply the Markov diagram to these special f of [6], and show how one can get the stronger results of [6] also with the methods of [4]. We always indicate the connection to the kneading invariant. Using f-expansions, in § 1 a shift space (X,σ) is constructed, whose nonwandering set Ω is investigated in § 2. One gets a decomposition of Ω splitting up the Markov diagram, which we consider as a 0-1-matrix M, into its irreducible submatrices. We show that the kneading invariant $\nu(f)$ of f is a sort of characteristic polynomial of M. The corresponding method of [6] is, to split up $\nu(f)$ into a product. § 3 considers the problem to bring the results about (X,σ) back to $([0,1],f)$.

§ 1 . f-expansions.

Let f be a continuous transformation on [0,1] with $f(0) = f(1) = 0$ and $f|[0,c]$ is strictly increasing and $f|[c,1]$ is strictly decreasing for some $c \in (0,1)$. We set $s = f(c)$ and $r = f^2(c)$. To avoid trivial cases, we assume $r < c < s$. It is easy to see that $[0,r) \cup (s,1]$ consists of wandering points and fixed points. In order to investigate the nonwandering set of f, we restrict f to $I := [r,s]$.

The f-expansion $\varphi: I \to \{0,1\}^N$ is defined by

$$\varphi(x) = \underline{x} = x_0 x_1 x_2 \ldots, \text{ where} \qquad (1.1)$$
$$x_i = 0 \text{ if } f^i(x) < c, \text{ and } x_i = 1 \text{ if } f^i(x) \geq c$$

1) Institut für Mathematik, Universität Wien, Strudlhofgasse 4, A-1090 Wien

We introduce an order relation $\triangleleft$ in $\{0,1\}^{\mathbb{N}}$. Let $b(x_0x_1\dots x_{m-1})$ denote the number of ones in the block $x_0x_1\dots x_{m-1}$. We define

$$\underline{x} = x_0x_1\dots \triangleleft \underline{y} = y_0y_1\dots,$$

if $\underline{x} = \underline{y}$ or, denoting by k the least coordinate i with $x_i \neq y_i$, if $b(x_0x_1\dots x_{k-1})$ is even and $x_k < y_k$, or if $b(x_0x_1\dots x_{k-1})$ is odd and $x_k > y_k$. Then it is not difficult to show that

$$x < y \Rightarrow \varphi(x) \triangleleft \varphi(y) \quad \text{for} \quad x,y \in I \tag{1.2}$$

Remark that $b(x_0x_1\dots x_{k-1})$ is the number of visits of $f^i(x)$ for $0 \le i \le k-1$ in $[c,s]$, where $x_0x_1x_2\dots = \varphi(x)$, and that $f|[c,s]$ inverts the ordering. If σ denotes the shift transformation on $\{0,1\}^{\mathbb{N}}$ given by $\sigma(x_0x_1x_2\dots) = x_1x_2\dots$, it follows from (1.1) that

$$\varphi \circ T = \sigma \circ \varphi \tag{1.3}$$

Next we define $\underline{e} = \lim_{t \uparrow s} \varphi(t) \in \{0,1\}^{\mathbb{N}}$ and

$$X = X_{\underline{e}} = \{\underline{x} \in \{0,1\}^{\mathbb{N}} : \sigma^k \underline{x} = x_k x_{k+1}\dots \triangleleft \underline{e} \text{ for } k \ge 0\} \tag{1.4}$$

If s is not periodic under f, it follows from (1.1) that $\underline{e} = \varphi(s)$. If s is periodic, change the definition (1.1) to $x_i = 0$ if $f^i(x) \le c$, and $x_i = 1$ if $f^i(x) > c$, if necessary, to get $\underline{e} = \varphi(s)$. Then we have

<u>Lemma 1</u>. (i) $X = \overline{\varphi(I)}$.

(ii) $X \setminus \varphi(I)$ is either $\{\underline{x} \in X : \sigma^k \underline{x} = 0\underline{e} \text{ for some } k \ge 0\}$ or $\{\underline{x} \in X : \sigma^k \underline{x} = 1\underline{e} \text{ for some } k \ge 0\}$.

The proof of this lemma is the same as that of theorem 1 and lemma 4 of [2].

In [6], kneading coordinates are used for the investigation of f. These are also a sort of expansion $\theta(x)$ of $x \in I$, written as power series, defined by

$$\theta(x) = \sum_{i=0}^{\infty} \theta_i(x) t^i,$$

where $\theta_i(x) = 1$ if f^i is order preserving near x, $\theta_i(x) = 0$ if $f^j(x) = c$ for some $j < i$, and $\theta_i(x) = -1$ if f^i is order reversing near x. As in (1.4), an important role is played by the expansion of c. The kneading invariant $\nu(f)$ is defined by $\nu(f) = \lim_{t \uparrow c} \theta(t)$. (cf. (2) of [6]).

We introduce the topology in X one gets by restricting the product topology of $\{0,1\}^{\mathbb{N}}$. σ is then continuous. There are two differences between (X,σ) and (I,f):

(i) Let $x \in I$ satisfy $f^k(x) = c$ for some $k \geq 0$. Then there are $\underline{x},\underline{y} \in X$ with $\underline{x} = \lim\limits_{t \downarrow x} \varphi(t)$ and $\underline{y} = \lim\limits_{t \uparrow x} \varphi(t)$. Neighbourhoods of $\underline{x}$ and $\underline{y}$ are intervals $[\underline{x},\underline{z})$ and $(\underline{w},\underline{y}]$ in X respectively, whereas the neighbourhoods of x are $(x-\varepsilon, x+\varepsilon)$. In all other points x, φ is continuous.

(ii) Certain intervals $J \subset I$ may be mapped to single points by φ.

§ 2. The structure of (X,σ).

In order to determine the topologically transitive closed σ-invariant subsets of (X,σ) we proceed as in [4]. We start with the following diagram, which gives a picture of X.

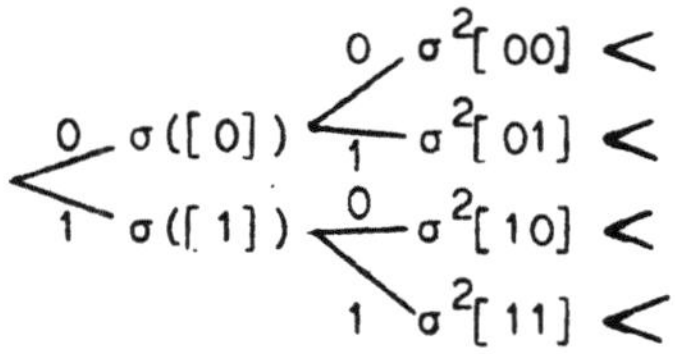

(2.1)

Here we set $[x_0x_1\ldots x_{k-1}] = \{\underline{y} \in X\colon y_i = x_i \text{ for } 0 \leq i \leq k-1\}$. The one-sided paths in (2.1), which do not lead to an empty set, represent the elements $x_0x_1\ldots \in X$, where $x_0x_1\ldots$ are the numbers of the arrows on the path. In theorem 1 below we shall compute (2.1) by induction, determining the successors of a set $\sigma^k([x_0\ldots x_{k-1}])$ by the formula

$$\sigma^{k+1}([x_0\ldots x_k]) = \sigma([x_k] \cap \sigma^k([x_0\ldots x_{k-1}])) \tag{2.2}$$

In order to do this, we define integers $r_1, r_2, \ldots$ $(r_i \geq 1)$ by induction. Set $r_1 = 1$. If $r_1, r_2, \ldots, r_k$ are defined, choose r_{k+1} such that

$$\begin{aligned} e_{r_1+\ldots+r_k+i} &= e_{i-1} \quad \text{for } 1 \leq i \leq r_{k+1}-1 \\ e_{r_1+\ldots+r_k+r_{k+1}} &\neq e_{r_{k+1}-1} \end{aligned} \tag{2.3}$$

If $r_k = \infty$ for some k, there are no r_j for $j > k$. We set $R_k = r_1 + \ldots + r_k$ and $R_0 = 0$. The following lemma is proved in [5] using only the fact that $\sigma^k\underline{e} \lhd \underline{e}$ for $k \geq 1$, i.e. $\underline{e} \in X$.

<u>Lemma 2</u>. If $r_i < \infty$ for $1 \leq i \leq k$ then

(i) $r_k = 1 + R_{Q(k)}$ for some $Q(k)$ with $0 \leq Q(k) \leq k-1$

(ii) $b(e_0e_1\ldots e_{R_k})$ is odd.

For a given n choose m such that $R_m < n \leq R_{m+1}$ and set

$$J_n = [\sigma^n\underline{e}, \sigma^{n-R_m-1}\underline{e}] \text{ or } [\sigma^{n-R_m-1}\underline{e}, \sigma^n\underline{e}],$$

depending on which of the two endpoints of J_n is smaller with respect to $\prec$. The square brackets denote closed intervals in $(X,\prec)$.

Theorem 1. (i) The J_n ($n\geq 1$) are all sets occurring in (2.1) and there are the arrows

$$J_n \to J_{n+1} \text{ for } n\geq 1 \text{ and } J_{R_k} \to J_{r_k} \text{ for } k\geq 1 \tag{2.4}$$

All arrows ending at J_{n+1} have the number e_n.

(ii) If $r_{m+1} = \infty$ for some m, then $\sigma^{R_m+1}\underline{e} = \underline{e}$ and $J_{2R_m+2} = J_{R_m+1} = \{\underline{e}\}$. Otherwise, all J_n are nontrivial intervals.

The oriented graph given by (2.4) is called the Markov diagram of f.

Proof. (i) Remark that $[0]$ is the closed interval $[\sigma\underline{e}, 0\underline{e}]$, whose endpoints begin both with 0. Hence $\sigma([0]) = [\sigma^2\underline{e},\underline{e}] = J_2$, because $r_1 = 1$. Similarly, $\sigma([1]) = \sigma([1\underline{e},\underline{e}]) = [\sigma\underline{e},\underline{e}] = J_1$. We proceed by induction using (2.2) to determine the successors of J_n.

Suppose first that $n \neq R_k$ for all $k\geq 0$. Then both endpoints of J_n begin with e_n by (2.3). Hence only $J_n \cap [e_n] = J_n$ is not empty, i.e. J_n has only the successor $\sigma(J_n) = J_{n+1}$. The arrow $J_n \to J_{n+1}$ has the number e_n.

Now suppose $n = R_k$ for some k. The endpoints of J_n are $e_{R_k}e_{R_k+1}\dots$ and $e_{r_k-1}e_{r_k}\dots$. By (2.3) we have $e_{R_k} \neq e_{r_k-1}$ and hence $J_n\cap[0] \neq \emptyset$ and $J_n\cap[1] \neq \emptyset$. Thus J_n has two successors. We have $\sigma(J_n\cap[e_{R_k}]) = [\sigma^{R_k+1}\underline{e},\underline{e}] = J_{R_k+1} = J_{n+1}$. Remark that $J_{n+1} = [\sigma^{R_k+1}\underline{e},\underline{e}]$, because $\underline{e}$ is the largest point of X. To get the other successor of J_n, we use (i) of lemma 2 and have

$$\sigma(J_n\cap[e_{r_k-1}]) = [\sigma^{r_k}\underline{e},\underline{e}] = [\sigma^{R_{Q(k)}+1}\underline{e},\underline{e}] = J_{R_{Q(k)}+1} = J_{r_k}$$

The arrow has the number e_{r_k-1}.

(ii) It follows from (2.3) that $\sigma^{R_m+1}\underline{e} = \underline{e}$ and from the definition of J_n, that $J_{R_m+1} = \{e\}$. $J_{2R_m+2} = \{\sigma^{R_m+1}\underline{e}\} = \{\underline{e}\}$. If $r_{m+1}<\infty$, then $\sigma^{R_m+1}\underline{e} \neq \underline{e}$ and J_{R_m+1} is nontrivial. □

The Markov diagram (2.4) satisfies lemmas 3 and 4 of [4], which are basic for all results of [4].

We consider the Markov diagram also as a 0-1-matrix M with index set $\mathfrak{D} = \mathbb{N}$ (if $r_{j+1} = \infty$, then $\mathfrak{D} = \{1,2,\dots,2R_j+1\}$; cf. (ii) of theorem 1)

$$M_{nm} = 1 \Leftrightarrow J_n \to J_m \text{ in } (2.4) \tag{2.5}$$

We split up M into irreducible submatrices. To this end we define integers $0 = k_0 < k_1 < \ldots < k_{u-1} < k_u = \infty$ ($u = \infty$ is possible) such that for $1 \le i < u$

$$r_j \ge R_{k_i}+1 \text{ for all } j \ge k_i+1 \tag{2.6}$$

By (i) of lemma 2, $r_{k_i+1} = R_{k_i}+1$ or ∞. For $1 \le i < u+1$ define ($R_\infty = \infty$)

$$\mathfrak{D}_i = \{m \in \mathfrak{D}:\ R_{k_{i-1}}+1 \le m < R_{k_i}+1\}\,;\ M^i = M|\mathfrak{D}_i \tag{2.7}$$

It follows from (2.4) and (2.5) that the M^i are the irreducible submatrices of M (cf. also (ii) of theorem 1), because by (2.6) no arrow leads from $\mathfrak{D}_i$ to $\mathfrak{D}_{i-1}$.

Lemma 3. The characteristic polynomial $C_m(\lambda)$ of the finite matrix $M|\{1,2,\ldots,R_m\}$ is given by

$$C_m(\lambda) = \sum_{k=0}^{R_m} a_k \lambda^{R_m-k}, \quad a_k = (-1)^{b(e_0\ldots e_{k-1})}, \quad a_0 = 1$$

Proof. This follows from the recursion formula $C_0(\lambda) = 1$ and

$$C_{m+1}(\lambda) = \lambda^{r_{m+1}} C_m(\lambda) - C_{Q(m+1)}(\lambda)$$

proved in lemma 5 of [5] and from (ii) of lemma 2 □

This lemma gives the connection between the kneading invariant ν and the 0-1-matrix M. It follows from the definition of ν that

$$\nu = \lim_{x \uparrow c} \theta(t) = \sum_{k=0}^{\infty} a_k t^k,$$

because f^k is order preserving near s, if $b(e_0\ldots e_{k-1})$ is even, and order reversing, if $b(e_0\ldots e_{k-1})$ is odd. If s is periodic, one has to consider $s-\varepsilon$ for small $\varepsilon > 0$, because also $\underline{e}$ is defined as a limit from below. We see, that the characteristic polynomials $C_m(\lambda)$ and the kneading invariant have the same coefficients.

Lemma 4. (i) For $0 < i < u$, the number $p_i := R_{k_i}+1$ is the greatest common divisor of $\{r_j:\ j \ge k_i+1,\ r_j < \infty\}$.

(ii) The matrix M^{i+1} has period p_i for $0 \le i < u$.

(iii) For $j > i$, p_i divides p_j.

Proof. (i) By (2.6) and (i) of lemma 2, we have $r_{k_i+1} = p_i$. Suppose we have shown that p_i divides r_j for $k_i + 1 \leq j < m$. By (i) of lemma 2 we get $r_m = (1 + R_{k_i}) + r_{k_i+1} + \ldots + r_{Q(m)}$, if $r_m < \infty$. Remark that (2.6) implies $Q(m) \geq k_i$. As $Q(m) \leq m-1$, it follows by induction that p_i divides r_m. (ii) and (iii) are immediate consequences of (i). Apply (ii) of theorem 1 to prove (ii), if $r_{k_i+1} = \infty$. □

We compare this result with theorem 1 of [6]. By (ii) of lemma 4, the characteristic polynomial of M^i, divided by its largest power, is of the form $P_i(\lambda^{-p_{i-1}})$, where P_i is a polynomial, if M^i is a finite matrix, and a power series, if M^i is infinite (cf. lemma 3). This second case happens only, if $i = u < \infty$ and $r_{k_{u-1}+1} < \infty$. Now it follows that the characteristic polynomial

$$C_j(\lambda) = \lambda^{R_j} P_1(\lambda^{-1}) P_2(\lambda^{-p_1}) \ldots P_i(\lambda^{-p_{i-1}})$$

where $j = k_i$. Remark that $p_m = R_{k_m}+1$ and R_{k_m} is the degree of $P_1(t) \ldots P_m(t^{p_{m-1}})$. This shows that this product can be extended to an infinite one. As the coefficients of C_j are exactly the first R_j+1 coefficients of ν, this gives the decomposition of theorem 1 of [6].

$$\nu = \prod_{i=1}^{u} P_i(t^{p_{i-1}}), \quad p_0 = 1.$$

One gets the different cases of this theorem, if one distinguishes $u < \infty$, $u = \infty$, $r_i < \infty$ for all i, $r_i = \infty$ for some i.

We conclude § 2 proving the results of [6] with the aid of the Markov diagram. We define $F_i = \bigcup\{J_m : m \geq p_{i-1}\} = \bigcup\{J_m : m \in \bigcup_{j \geq i} \mathfrak{D}_j\}$. Then $F_{i+1} \subset F_i$ and $\sigma(F_i) \subset F_i$ by (2.6), because $\sigma(J_n)$ is the union of the successors of J_n in the Markov diagram (cf (2.2)). Lemma 1 of [4] says that F_i is a finite union of intervals. For the special transformations we consider in this paper we have a stronger result and an easier proof.

Lemma 5. (i) For $p_i \leq j \leq 2p_i-2$, σ maps J_j monotonically to J_{j+1}. $J_{2p_i-1} \cap [0]$ and $J_{2p_i-1} \cap [1]$ are mapped monotonically to J_{p_i} and J_{2p_i} or vice versa. $J_{2p_i} \subset J_{p_i}$, hence $\sigma(J_{2p_i-1}) = J_{p_i}$.

(ii) $F_{i+1} = \bigcup\{J_m : p_i \leq m \leq 2p_i - 1\}$ for $0 \leq i < u$.

Proof. (i) Remark that $p_i = R_{k_i}+1$ and $2p_i - 1 = R_{k_i} + R_{k_i} + 1 = R_{k_i} + r_{k_i+1} = R_{k_i+1}$. Hence (i) follows from the proof of theorem 1, except that $J_{2p_i} \subset J_{p_i}$.

If $r_{k_i+1} = \infty$, then we have by (ii) of theorem 1 that $J_{p_i} = J_{2p_i} = \{\underline{e}\}$. Hence it suffices to consider $r_{k_i+1} < \infty$. Then we have that

$$J_{p_i} = [\sigma^{p_i}\underline{e}, \underline{e}] \text{ and } J_{2p_i} = [\sigma^{2p_i}\underline{e}, \underline{e}],$$

because $p_i = R_{k_i}+1$, $2p_i = R_{k_i+1}+1$, and $\underline{e}$ is the largest element of X. If $r_j = r_{k_i+1} = p_i$ for all $j \geq k_i+1$, we have $\sigma^{p_i}\underline{e} = \sigma^{2p_i}\underline{e}$ by (2.3), hence $J_{p_i} = J_{2p_i}$. If $r_m > r_{k_i+1} = p_i$ for some $m > k_i+1$ (choose m minimal), it follows that $\sigma^{p_i}\underline{e} \triangleright \sigma^{2p_i}\underline{e}$, because $e_j = e_{p_i+j}$ for $p_i \leq j < R_{m-1}$ and $e_{R_{m-1}} \neq e_{R_{m-1}+p_i}$ by (2.3). Consider the two cases $e_{R_{m-1}} = 0$ and $e_{R_{m-1}} = 1$ and use (ii) of lemma 2.

(ii) is an easy consequence of (i), which implies by induction that $J_{m+np_i} \subset J_m$, where $p_i \leq m \leq 2p_i - 1$, because $\sigma(J_n)$ is the union of the successors of J_n in the Markov diagram. $\square$

As φ is order preserving, this lemma implies that f maps the interval $\varphi^{-1}(J_j)$ monotonically to the interval $\varphi^{-1}(J_{j+1})$ for $p_i \leq j \leq 2p_i - 2$, and that the interval $\varphi^{-1}(J_{2p_i-1})$ is folded and mapped to $\varphi^{-1}(J_{p_i})$. Hence f^{p_i} restricted to one of the intervals $\varphi^{-1}(J_j)$, $p_i \leq j \leq 2p_i - 1$, is again a transformation of the form considered in this paper. Its Markov diagram is determined by the numbers $\bar{r}_m = r_{m+k_i}/p_i$ for $m \geq 1$ (cf. Lemma 4).

Using the σ-invariant sets F_i it is easy to decompose the nonwandering set Ω of (X,σ). We set $F_{u+1} = \emptyset$ and define for $1 \leq i < u+1$

$$\Omega_i = \bigcap_{m=0}^{\infty} \overline{\sigma^{-m}(F_i \setminus F_{i+1})}, \quad W_i = (F_i \setminus F_{i+1}) \setminus \Omega_i .$$

One sees easily that Ω_i is closed and σ-invariant. It is proved in [4], that Ω_i is topologically transitive, and hence $\Omega_i \subset \Omega$. We shall show that $W_i \subset X \setminus \Omega$.

<u>Lemma 6</u>. (i) bd $F_{i+1} \subset \{\sigma^m \underline{e}: 0 \le m \le 2p_i - 1\}$; bd $F_i \subset$ bd F_{i+1}.
(ii) For $i < j$, $\Omega_i \cap F_j \neq \emptyset$ implies $u < \infty$ and $j = i+1 = u$. Furthermore $\Omega_{u-1} \cap F_u = \Omega_{u-1} \cap \Omega_u$ is then the periodic orbit $\{\sigma^m \underline{e}: p_i \le m \le 2p_i - 1\}$.

<u>Proof</u>. The first assertion of (i) follows from (ii) of lemma 5, because J_m has endpoints $\sigma^m \underline{e}$ and $\sigma^{m-p_i} \underline{e}$ for $p_i \le m \le 2p_i - 1$. To show the second assertion, let $\underline{x} \in$ bd F_i, i.e. $\underline{x} = \sigma^m \underline{e}$ with $0 \le m \le 2p_{i-1} - 1$. As $p_i \ge 2p_{i-1}$ (cf. (2.6)), $\underline{x}$ is also an endpoint of $J_{p_i+m} \subset F_{i+1}$. If $\underline{x} \in$ int F_{i+1}, then $\underline{x} \in$ int F_i, because $F_{i+1} \subset F_i$, a contradiction. Hence $\underline{x} \in$ bd F_{i+1}.

To show (ii) remark that $\Omega_i \subset \overline{F_i \setminus F_{i+1}}$ and $F_j \subset F_{i+1}$. If $\underline{x} \in \Omega_i \cap F_j$, then $\underline{x} \in$ bd F_{i+1} and hence $\underline{x} = \sigma^m \underline{e}$ for some m with $0 \le m \le 2p_i - 1$ by (i). As Ω_i and F_j are σ-invariant, the same is true for $\Omega_i \cap F_j$, and hence $\sigma^k \underline{x} \in \Omega_i \cap F_j$. As for $\underline{x}$, we get also $\sigma^k \underline{x} \in$ bd F_{i+1}. By lemma 5, we have $J_{2p_i} \subset J_{p_i}$. Hence $\sigma^{2p_i - m} \underline{x} = \sigma^{2p_i} \underline{e} \in$ bd F_{i+1} only, if $J_{p_i} = J_{2p_i}$, i.e. $\sigma^{2p_i} \underline{e} = \sigma^{p_i} \underline{e}$. This implies $r_j = r_{k_i+1}$ for all $j \ge k_i + 1$ or $r_{k_i} + 1 = \infty$. In both cases we have $u = i+1 < \infty$ and $j = u$, as $i < j \le u$. Furthermore $\Omega_u = F_u$ and $\Omega_{u-1} \cap F_u$ is the periodic orbit $\{\sigma^m \underline{e}: p_i \le m \le 2p_i - 1\}$. □

<u>Lemma 7</u>. $W_i \subset X \setminus \Omega$.

<u>Proof</u>. It follows from (i) of lemma 6 that $F_i \setminus F_{i+1}$ is open, since bd$(F_i \setminus F_{i+1}) \subset$ bd $F_{i+1} \subset F_{i+1}$. If $\underline{x} \in W_i$, then $\underline{x} \notin \Omega_i$ and there is an m with $\underline{x} \notin \overline{\sigma^{-m}(F_i \setminus F_{i+1})}$. This set is closed. Hence $\underline{x}$ has a neighbourhood $V = [x_0 \ldots x_{n-1}]$ with $n \ge m$, such that $V \subset F_i \setminus F_{i+1}$ and $V \cap \sigma^{-m}(F_i \setminus F_{i+1}) = \emptyset$. As $n \ge m$ and $\sigma(F_i) \subset F_i$, this gives $\sigma^m V \subset F_{i+1}$. Since $\sigma(F_{i+1}) \subset F_{i+1}$ we get $\sigma^j V \cap V = \emptyset$ for $j \ge m$. Making n larger, if necessary, we get also $\sigma^j V \cap V = \emptyset$ for $1 \le j < m$. Hence $\underline{x} \notin \Omega$. □

Remark that $F_1 = X$, because $J_1 = [\sigma \underline{e}, \underline{e}] = X \subset F_1$. Hence for $u < \infty$, the sets $F_i \setminus F_{i+1}$ for $1 \le i \le u$ form a partition of X. Lemma 7 implies that $\Omega = \bigcup_{i=1}^{u} \Omega_i$.

If $u = \infty$, the sets $F_i \setminus F_{i+1}$ for $1 \leq i < \infty$ and $\Omega_\infty = \bigcap_{i=1}^{\infty} F_i$ form a partition of X. Ω_∞ is nonempty as an intersection of a decreasing sequence of compact sets. It is σ-invariant, because F_i is, and $\Omega_\infty \cap \Omega_i = \emptyset$ by (ii) of lemma 6. Using the fact that the onesided paths in the Markov diagram represent the elements of X, one can show that $\{\sigma^m \underline{e}: m \geq 0\}$ is dense in Ω_∞, i.e. Ω_∞ is topologically transitive and $\Omega_\infty \subset \Omega$. By lemma 7, we have $\Omega = \bigcup_{1 \leq i \leq \infty} \Omega_i$.

The following theorem summarizes the results about Ω and contains also some results which follow from [4], [3] and [2].

<u>Theorem 2</u>. $\Omega = \bigcup_{1 \leq i \leq u} \Omega_i$ for some u with $1 \leq u \leq \infty$.

(i) Ω_i is topologically transitive, closed and σ-invariant.

(ii) $\Omega_i \cap \Omega_j$ is empty or finite.

(iii) For $i < u$, Ω_i is a periodic orbit or a Cantorset and a finite type subshift. For $u < \infty$, Ω_u is a periodic orbit, or a finite union of intervals. For $u = \infty$, Ω_u is a Cantorset, which contains no periodic points and has topological entropy zero.

(iv) For $i < \infty$, $h_{top}(\Omega_i) = \log r(M^i)$, $r(M^i)$ is the spectral radius of M^i. If $i < j$ and $r(M^i) > 1$, $r(M^j) > 1$, then $r(M^i) > r(M^j)$.

(v) For $i < \infty$, set $Y_j = \Omega_i \cap J_j$ for $p_{i-1} \leq j \leq 2p_{i-1}-1$. Then $\sigma(Y_j) \subset Y_{j+1}$ for $p_{i-1} \leq j \leq 2p_{i-1}-2$ and $\sigma(Y_{2p_{i-1}-1}) \subset Y_{p_{i-1}}$. $\sigma^{p_i} | Y_j$ is topologically mixing.

§ 3. The nonwandering set of (I,f).

If one wants to bring the results of § 2 from (X,σ) back to (I,f), one has to consider the two problems mentioned at the end of § 1. We denote the nonwandering set of (I,f) by L.

If $x \in I$ is not an inverse image of c under f and x is not in a nontrivial interval $\subset I$ mapped to a single point by φ, then one sees easily that $x \in L$, if and only if $\varphi(x) \in \Omega$, because the neighbourhoods of x and $\varphi(x)$ correspond to each other via φ by (1.2).

We consider an $x \in I$ with $f^k(x) = c$ for some $k \geq 0$. We show first that

$$\underline{e} \in \operatorname{int} F_i \text{ for } 1 \leq i < u+1 \tag{3.1}$$

This is clear, if $J_{p_i} = [\underline{z},\underline{e}] \subset F_i$ satisfies $\underline{z} \neq \underline{e}$. If $J_{p_i} = \{\underline{e}\}$, then $r_{m+1} = \infty$ for $m = k_i$ by (ii) of theorem 1. It follows then from (2.3) and (ii) of lemma 2 that $[e_0 \ldots e_{R_m+1}] = \{\underline{e}\}$, which is an open set. This gives (3.1). It follows from (3.1) that

$$s \in \operatorname{int} \varphi^{-1}(F_i) \text{ for } 1 \leq i < u+1 \tag{3.2}$$

because, due to the definition $\underline{e} = \lim_{t \uparrow s} \varphi(t) = \varphi(s)$, φ is continuous in s. In the case, where $\{\underline{e}\}$ is open, $\varphi^{-1}\{\underline{e}\}$ is a nontrivial interval in I.

<u>Lemma 8</u>. Let $x \in I$ be such that $f^k(x) = c$ for some $k \geq 0$, $x \neq r,s$. Let $\underline{x},\underline{y} \in X$ be $\lim_{t \uparrow x} \varphi(t)$ and $\lim_{t \downarrow x} \varphi(t)$ respectively. (For $x = r$ or s, either $\underline{x}$ or $\underline{y}$ does not exist and φ is continuous in r and s.)

(i) If $\underline{x},\underline{y} \notin F_i$ for some i, then $\underline{x},\underline{y} \notin \Omega$ and $x \notin \overline{L}$.

(ii) If either $\underline{x}$ or $\underline{y}$ is in $\cap F_i = \Omega_u$, then $x \in \overline{\varphi^{-1}(\Omega_u)}$, i.e. $x \in L$ or x is in the interior of an interval mapped to a single point by φ.

<u>Proof</u>. (i) By (3.1) and $\sigma^{k+1}\underline{x} = \sigma^{k+1}\underline{y} = \underline{e}$, we find neighbourhoods U of $\underline{x}$ and V of $\underline{y}$, such that $\sigma^{k+1} U$ and $\sigma^{k+1} V$ are in F_i. As F_i is σ-invariant, we get as in the proof of lemma 7 that $\underline{x},\underline{y} \notin \Omega$. The same proof works for x using (3.2) and gives $x \notin L$.

(ii) By topological transitivity of Ω_u and (1.2) we get the topological transitivity of $\overline{\varphi^{-1}(\Omega_u)}$, if we take away the interiors of intervals $\subset \overline{\varphi^{-1}(\Omega_u)}$ mapped to single points by φ. That $x \in \overline{\varphi^{-1}(\Omega_u)}$ follows, because $\underline{x}$ or $\underline{y}$ is in Ω_u.

Now we consider a nontrivial interval $J \subset I$ satisfying $\varphi(J) = \{\underline{x}\}$ for some $\underline{x} \in X$. If $\underline{x}$ is not periodic, then $J \subset I \setminus L$. If $\sigma^p\underline{x} = \underline{x}$ for some $\underline{x}$, then $f^p(J) \subset J$ and f^p/J is monotone. Hence $J \cap L$ consists only of periodic points. This problem is thus easy, if one knows the $\underline{x} \in X$ with $\varphi^{-1}(\{\underline{x}\})$ a nontrivial interval. This is solved for a class of transformations including $x \mapsto ax(1-x)$ for $a \leq 4$ in [1]. Such $\underline{x}$ exist only if $\sigma^p\underline{e} = \underline{e}$ for some p and then these $\underline{x}$ are exactly the inverse images of $\underline{e}$. For $x \mapsto ax(1-x)$ (more generally for the transformations considered in [1]) one gets from theorem 2 and the results of [1] the following.

<u>Theorem 3.</u> $L = \bigcup_{1 \le i \le u} L_i$ for some u with $1 \le u \le \infty$.

(i) L_i is topologically transitive, closed and f-invariant.

(ii) For $i < j < u$, $L_i \cap L_j = \emptyset$. For $i < u$, L_i is a Cantorset or a periodic orbit and has Lebesguemeasure 0. (L_i, f) is topologically isomorphic to a finite type subshift.

(iii) If $u < \infty$, then L_u is a periodic orbit or a finite union of intervals. For $i < u-1$, $L_i \cap L_u = \emptyset$ and $L_{u-1} \cap L_u$ is empty or finite.

If $u = \infty$, then L_u contains no periodic orbit and is the closure of the orbit of c. For $i < u$, $L_i \cap L_u = \emptyset$.

References.

[1] J. Guckenheimer: Sensitive dependence to initial conditions for one dimensional maps. Comm.Math.Ph. <u>70</u> (1979), 133-160.

[2] F. Hofbauer: On intrinsic ergodicity of piecewise monotonic transformations with positive entropy. Israel J. Math. <u>24</u> (1979), 213-237.

[3] F. Hofbauer: On intrinsic ergodicity of piecewise monotonic transformations with positive entropy II. Israel J. Math. <u>38</u> (1981), 107-115.

[4] F. Hofbauer: The structure of piecewise monotonic transformations. Ergod.Th. and Dynam.Sys. (1981).

[5] F. Hofbauer: The topological entropy of the transformation $x \longrightarrow ax(1-x)$. Mh. Math. <u>90</u> (1980), 117-141.

[6] L. Jonker, D. Rand: Bifurcations in one dimension. Invent. math. <u>62</u> (1981), 347-365.

SUFFICIENT STATISTICS OF CLASSICAL AND QUANTUM DYNAMICAL SYSTEMS

Roman S. Ingarden[1)]

Summary. It is shown that the theory of classical and quantum dynamical systems can be formulated in a category-theoretical invariant way, i.e. as a quasi-geometry (in analogy to F.Klein's programme of expressing geometry as a group-theoretical invariant theory). This idea is a reformulation of N.Chentsov's and N.Morse-R.Sacksteder's ideas from the theory of statistical decision rules, as well as a further development of R.A.Fisher's idea of a sufficient statistic. Two types of dynamical systems are considered: deterministic and stochastic ones, although both are statistical (the first two terms concern the type of the time evolution law, the latter the character of states). The stochastic dynamical systems are defined by means of Markov morphisms (transition probabilities). The evolution law is not necessarily linear in the considered state space.

As M. Fisz remarked, [8] p. 353, "the term statistic belongs to the most ambiguous ones". In the sense in which it will be used here, and in which it is now used in mathematical statistics, for a sample space X with elements x and a σ-algebra of subsets S, it corresponds to "a partitioning of X into sets of equivalent x's" , [14] p. 18, cf. also [15] p. 6-7,

$$X = \bigcup_i E_i \ , \ E_i \in S \ , \ E_i \cap E_j = \emptyset \ , \ i \neq j \ , \ i,j=1,2,\ldots \qquad (1)$$

In physics such a process is called coarse graining. We have the theorem that if (X,S,m_k) , $k=1,2$, are two probability spaces, and three probability measures on S, m_1, m_2, m , are mutually absolutely continuous, $f_k(x) = dm_k/dm$, then

1) Institute of Physics, N.Copernicus University, Grudziadzka 5, 87-100 Toruń, Poland.

$$I_X(1:2) = \int_X f_1(x) \ln\left(\frac{f_1(x)}{f_2(x)}\right) m(dx) \geq \sum_i m_1(E_i) \ln\left(\frac{m_1(E_i)}{m_2(E_i)}\right) =$$

$$= I_Y(1:2) \ , \tag{2}$$

where $Y = \{E_i : i=1,2,\dots\}$ may be considered as a second sample space, with equality in (2) iff

$$\frac{f_1(x)}{f_2(x)} = \frac{m_1(E_i)}{m_2(E_i)} \, [m] \ , \ x \in E_i \ , \ i=1,2,\dots \tag{3}$$

$[m]$ meaning "modulo measure m" , cf. [14] p. 16, Corollary 3.2. The quantities $I_X(1:2)$ and $I_Y(1:2)$ are relative informations (relative entropies) of measure m_1 with respect to measure m_2 in the respective sample spaces X and Y. If we denote by T=T(Y) the power set of the discrete set Y and

$$p_k = \{ m_k(E_i) = \int_{E_i} f_k(x) \, m(dx) \ : \ i=1,2,\dots \} \ , \ k=1,2, \tag{4}$$

we obtain the map

$$F : (X,S,m_k) \longrightarrow (Y,T,p_k) \tag{5}$$

such that, iff condition (3) is satisfied , the information

$$I_X(1:2) = I_Y(1:2) \tag{6}$$

is invariant under this map, otherwise being diminished. We say that the statistic (5) is a sufficient statistic with respect to measure m_1 and m_2 iff the condition (3) or (6) is satisfied . This definition is consistent with the original idea of sufficiency introduced by the eminent British statistician R.A.Fisher who expressed it by saying: " the statistic chosen should summarize the whole of the relevant information supplied by the sample", [9] p. 316.

More generally, let (X,S) and (Y,T) be two measurable spaces and $x \in X$, $y \in Y$. Then a map

$$F : X \longrightarrow Y \quad \text{or} \quad F : x \longmapsto y = F(x) \tag{7}$$

is said to be a statistic or statistic homomorphism (also random variable in the general sense, while a random variable usually considered is a numerical one, when $Y=R^1$ and T is the set of all Borel sets on R^1) iff F is measurable, i.e.,

$$F^{-1}(T) \subseteq S \,, \tag{8}$$

cf. [16] Chap. 2,1 , [5] p. 24. An indexed set of probability measures (statistical states or simply <u>states</u>) on (X,S), called <u>state space</u>,

$$P = \{ m_u : u \in U \} \tag{9}$$

(index or parameter u can be multidimensional, for simplicity we shall also say "state u" and "state space U") together with (X,S) is said to be a (<u>dominated</u>) <u>statistical structure</u> (X,S,P) iff all $m_u \in P$ are mutually absolutely continuous and also mutually absolutely continuous with respect to an arbitrary, but fixed σ-finite measure m on (X,S),

$$m_u \equiv m_v \equiv m \,, \quad u,v \in U, \tag{10}$$

cf. [4] Chap. I§1. Using (7) we define on (Y,T)

$$\begin{aligned} & n_u(A) = m_u(F^{-1}(A)) \,, \quad n(A) = m(F^{-1}(A)) \,, \quad u \in U, \; A \in T \,, \\ & Q = \{ n_u : u \in U \} \,. \end{aligned} \tag{11}$$

Then, cf. [14] p.19,

$$n_u \equiv n_v \equiv n \,, \quad u,v \in U \tag{12}$$

and (Y,T,Q) forms a dominated statistical structure. Defining the probability densities as Radon-Nikodym derivatives by

$$\begin{aligned} f_u(x) &= \frac{dm_u}{dm} = \frac{m_u(dx)}{m(dx)} \quad x \in X \,, \\ g_u(y) &= \frac{dn_u}{dn} = \frac{n_u(dx)}{n(dx)} \,, \quad y \in Y, \quad u \in U \end{aligned} \tag{13}$$

we obtain in analogy to (2) and (3)

$$I_X(u:v) = \int_X f_u(x) \ln\left(\frac{f_u(x)}{f_v(x)}\right) m(dx) = \int_Y g_u(y) \ln\left(\frac{g_u(y)}{g_v(y)}\right) n(dy) =$$

$$= I_Y(u:v) \,, \quad u,v \in U \,, \tag{14}$$

iff

$$\frac{f_u(x)}{f_v(x)} = \frac{g_u(F(x))}{g_v(F(x))} \; [m] \,, \quad x \in X, \; u,v \in U, \tag{15}$$

otherwise $I_X(u:v) > I_Y(u:v)$. We say that statistic F is a <u>sufficient statistic</u> with respect to the statistical structure (X,S,P) iff (15) is satisfied . Then (X,S,P) can be called <u>informationally homomorphic</u>

with (Y,T,Q) and map F an <u>information</u> <u>homomorphism</u>. Iff F is a bijection (on and 1-to-1), then besides of (7) and (15) we have also the symmetrical expressions

$$F^{-1} : Y \longrightarrow X \quad \text{or} \quad F^{-1} : y \longmapsto x = F^{-1}(y) \;, \tag{16}$$

$$\frac{f_u(F^{-1}(y))}{f_v(F^{-1}(y))} = \frac{g_u(y)}{g_v(y)} \;[n] \;, \; y \in Y, \; u,v \in U \;, \tag{17}$$

and we call F an <u>equivalent</u> <u>statistic</u> or <u>information</u> <u>isomorphism</u>, and (X,S,P) <u>information</u> <u>isomorphic</u> with (Y,T,Q) , cf. [17] . We may also say then that (X,S,P) and (Y,T,Q) are two informationally equivalent representations of our "abstract" state space U. In this abstract space U the <u>information</u> <u>distance</u> (or <u>quasi-distance</u> or <u>divergence</u>) of state u from state v can be defined

$$D(u:v) = (2I_X(u:v))^{1/2} = (2I_Y(u:v))^{1/2} \geqslant 0 \;, \; u,v \in U, \tag{18}$$

in general asymmetric, $D(u:v) \neq D(v:u)$, but such that

$$D(u:v) = 0 \quad \text{iff} \quad u=v, \; u,v \in U, \tag{19}$$

cf. [12] where further literature is given. For continuous parameters u,v, and $v = u+du$, $du \to 0$, i.e., locally, in the small, the quasi-metric D(u:v) goes over not only into a "normal", symmetric Fréchet's metric $d(u,v)=d(v,u)$, but also into a Riemannian metric

$$ds^2 = D^2(u:u+du) = d^2(u,u+du) = g_{ij}(u)du^i du^j \;, \; u \in U, \tag{20}$$

called in this case <u>Fisher's</u> <u>information</u> [9] , [14] , [5] .

To get more geometrical features of our construction, we extend the statistical structures as considered above by taking their <u>convex</u> <u>hull</u> in the sense of statistical mixing of states. Namely, we add to P (9) all convex combinations of any its elements with densities $f_u(x)$, $f_v(x)$, $u,v \in U$, or all their <u>mixed</u> <u>states</u>

$$\langle p,u,v \rangle = f^p_{uv}(x) = p\, f_u(x) + (1-p)\, f_v(x) \;, \quad 0 < p < 1 \;, \tag{21}$$

extending, if necessary, the corresponding parameter space U. Thus extended P will be called the <u>statistical</u> <u>figure</u> of P and denoted by P^c. We say that P is a statistical figure iff $P = P^c$. A statistical figure is <u>complete</u> or <u>full</u> iff it contains all possible states on (X,S). In statistical figures we have the important property of joint convexity of D-function, cf. [12], [20],

$$D^2(\langle p,u,v\rangle:\langle p,s,t\rangle) \leq pD^2(u:s)+(1-p)D^2(v:t) \quad , \quad u,v,s,t \in U \quad . \qquad (22)$$

By means of information distance D(u:v) information neighbourhoods (information spheres)

$$V(u;r) = \{v \in U \ : \ D(v:u) < r\} \quad , \quad u \in U \quad , \quad 0 < r \leq \infty \qquad (23)$$

can be defined. I.Csiszár [6] has shown that, in general, these neighbourhoods do not define a topology, but only a more general Fréchet (V)-space [19]. However, it can be easily shown [6] that in any state space P there exists a metric topology defined by the variation distance

$$R_X(u,v) = R_X(v,u) = \int_X |f_u(x)-f_v(x)| \, m(dx) \geq 0 \quad , \quad u,v \in U, \qquad (24)$$

which is, of course, a Fréchet metric. We have

Theorem 1. If convex statistical structures (X,S,P) and (Y,T,Q) are statistically isomorphic, then.

$$R_X(u,v) = R_Y(u,v) = R(u,v) \quad , \quad u,v \in U. \qquad (25)$$

Proof.

$$R_X(u,v)-R_Y(u,v) = \int_X f_v(x) \left| \frac{f_u(x)}{f_v(x)} - 1 \right| m(dx) - \int_Y g_v(y) \left| \frac{g_u(y)}{g_v(y)} - 1 \right| n(dy) =$$

$$= \int_X m_v(dx) \left| \frac{f_u(x)}{f_v(x)} - 1 \right| - \int_Y n_v(dy) \left| \frac{g_u(y)}{g_v(y)} - 1 \right| =$$

$$= \int_X m_v(dx) \left(\left| \frac{f_u(x)}{f_v(x)} - 1 \right| - \left| \frac{g_u(F(x))}{g_v(F(x))} - 1 \right| \right) = 0 \quad , \quad u,v \in U, \qquad (26)$$

the latter equalities following from (11),(13) and (15). □

We remark that information distance D(u:v) and variation distance R(u,v) can be imbedded into the family of Csiszár's f-divergencies [6]

$$I_{Xf}(u:v) = \int_X f_v(x) \, f\left(\frac{f_u(x)}{f_v(x)}\right) m(dx) \qquad (27)$$

where f is an arbitrary convex function on the real half axis $R^1_+ = (0,+\infty)$. For D and R we have, respectively,

$\frac{1}{2} D^2$: $f(w) = w \ln w$, R : $f(w) = |w-1|$, $w \in R^1_+$. (28)

Theorem 2. If convex statistical structures (X,S,P) and (Y,T,Q) are statistically isomorphic, and f is an arbitrary convex function on R^1_+ , then

$$I_{Xf}(u:v) = I_{Yf}(u:v) = I_f(u:v) , \quad u,v \in U. \tag{29}$$

Proof. Similarly as in (26) we have

$$I_{Xf}(u:v) - I_{Yf}(u:v) = \int_X m_v(dx)\left(f\left(\frac{f_u(x)}{f_v(x)}\right) - f\left(\frac{g_u(F(x))}{g_v(F(x))}\right)\right) = 0 , \quad u,v \in U. \ \square \tag{30}$$

It is well-known that the partial order relation of "mixing-enhancement" between states, so well investigated by A.Uhlmann and his school [1] , is uniquely connected with the family of all f-divergencies [20] ,[1] . Therefore, also this relation is independent of the choice of equivalent (or sufficient) statistic (representation).
All Csiszâr's f-divergencies (27) can be considered as some generalized relative informations (f-informations or f-entropies), but only for two particular f (28) they have simple mathematical properties and simple interpretation in physics and information theory [2] , [12] . The first is connected with the Boltzmann-Shannon-von Neumann entropy which for a single elementary event E with probability p(E) can be defined as <u>absolute</u> and <u>relative</u> [1)] <u>entropy</u>, respectively,

$$J(E) = -\ln p(E) , \quad I(E) = -p(E)\ln p(E) , \quad 0 \leq p(E) \leq 1. \tag{31}$$

The second special f-divergency, R, can be connected with the concept of a <u>content</u>, <u>absolute</u> and <u>relative</u> , respectively, of E

$$B(E) = |p(E)-1| = 1 - p(E) , \quad C(E) = p(E)(1-p(E)) , \quad 0 \leq p(E) \leq 1, \tag{32}$$

as defined by R.Carnap and Y.Bar-Hillel [3] , cf. also [12] .

---

1) "Relative" is meant here with respect to other events, while in the case of the relative entropy (2) "relative" is understood with respect to another state.

Then (<u>mean</u>) <u>entropies</u> can be defined for $p = \{ p_i = p(E_i): i=1,\dots, n, p_i \geqslant 0 , \sum_i p_i = 1 \}$

$$I(p) = - \sum_{i=1}^{n} p_i \ln p_i \geqslant 0 \quad , \quad C(p) = \sum_{i=1}^{n} p_i(1-p_i) = 1 - O(p) \geqslant 0 \qquad (33)$$

where $O(p)$ is the <u>Onicescu information</u> or <u>information energy</u> [18], [12] ,

$$0 < O(p) = \sum_{i=1}^{n} p_i^2 \leqslant 1 \ . \qquad (34)$$

The variation (or content) distance R is connected with the linear mixing operation (21) and defines an affine geometry with straight lines (segments) between states (points) as geodetics. The information distance D defines another "logarithmic geometry" (or quasi-geometry) with "geodesics" fixed by the so-called <u>geodesic mean</u> or <u>multiplicative mixing</u> [5]

$$[p,u,v] = f_{uv,p}(x) = Z^{-1}(p) [f_u(x)]^p [f_v(x)]^{1-p} \quad , \ 0 \leqslant p \leqslant 1, \qquad (35)$$

$$0 < Z(p) = \int_X [f_u(x)]^p [f_v(x)]^{1-p} \, m(dx) \leqslant 1 \ , \qquad (36)$$

or

$$\ln[p,u,v] = p \ln f_u(x) + (1-p) \ln f_v(x) - \ln Z(p) \ . \qquad (37)$$

Introducing the random variable (or sufficient statistic)

$$q(x) = \ln\left(\frac{f_v(x)}{f_u(x)}\right) \quad , \ x \in X \ , \qquad (38)$$

we can rewrite (37) in the form of the so-called <u>exponential family</u> [14], [5]

$$[p,u,v] = f_{uv,p}(x) = Z^{-1}(p) f_v(x) \exp[- p \, q(x)] \quad , \ 0 \leqslant p \leqslant 1. \qquad (39)$$

Here, as in (21), parameter p can be also extended outside of the interval [0,1] . Then we obtain a similarity to the Gibbs families of states in statistical thermodynamics [13], [10], [12], and exactly the families of solutions obtained in information-theoretical testing of statistical hypotheses [14] , [5]. In thermodynamics $q(x)$ is usually interpreted as energy, and p as the inverse temperature, while in testing of hypotheses $q(x)$ is called a directional sufficient statistic, and $-p$ a canonical affine parameter [5].

Let us assume that state space U is convex and closed in R-topology. Then it can be shown [7] that if E is a convex set of states closed in R-topology, there exists the unique state $v \in E$ such that

$$D(v:u) = \min_{w \in E} D(w:u) \tag{40}$$

called the <u>information projection</u> (I-<u>projection</u>) of u on E. If according to Csiszár [7] we define the <u>tangent hyperplane</u> to the information sphere $V(u;r)$, $r=D(v:u)<\infty$, at v as the set of states w satisfying the equation

$$\int_X f_w(x)\ln\left(\frac{f_v(x)}{f_u(x)}\right)m(dx) = r \ , \quad w \in U \ , \tag{41}$$

then it can be shown [7] that

$$D^2(w:u) = d^2(w:v) + D^2(v:u) \ , \tag{42}$$

cf. also [5],[12]. Property (42) can be interpreted as the <u>Pythagoras theorem</u> (actually <u>directed Pythagoras theorem</u>) of information geometry (D-geometry) since the triple (w,v,u) can be considered as a directed rectangular triangle with the right angle by v being the I-projection of u on the tangent hyperplane at v. Except of (18),(19),(22) and (40),(42) is the fifth fundamental property of D-geometry.

The class of all measurable spaces (X,S), as the class of objects, together with the class of all statistics F (7) between measurable spaces, as the class of morphisms, form the <u>category of measurable spaces</u> which we denote by MEAS = (Obj MEAS, Mor MEAS), cf. [5]p. 83. We have

<u>Theorem 3.</u> Let (X,S,P) be a convex statistical structure closed in R-topology. This structure generates a category, being a subcategory of MEAS, namely, $\mathcal{C}(X,S,P)=(\mathrm{Obj}\ \mathcal{C}(X,S,P),\ \mathrm{Mor}\ \mathcal{C}(X,S,P))$ where the elements of

$$\mathrm{Obj}\ \mathcal{C}(X,S,P) \subseteq \mathrm{Obj\ MEAS}\ , \quad \mathrm{Mor}\ \mathcal{C}(X,S,P) \subseteq \mathrm{Mor\ MEAS} \tag{43}$$

are selected by condition (15). All the morphisms $\in \mathrm{Mor}\ \mathcal{C}(X,S,P)$ are sufficient statistics (information homomorphisms). Iff except of (15) we assume also (16), i.e., that all the morphisms are equivalent statistics (information isomorphisms), we obtain the category

$\mathscr{C}_o(X,S,P)$ being a subcategory of $\mathscr{C}(X,S,P)$. Iff P is complete (full), $P=P(X,S)$, and we can denote the mentioned categories as $\mathscr{C}(X,S)$ and $\mathscr{C}_o(X,S)$, respectively.

The proof is obvious, as the proof of

Theorem 4. By a morphism $F \in$ Mor MEAS probability measure on a measurable space (X,S) is invariant in the sense of (11), but relative information of two states on (X,S) in general increases. If we interpret F as a movement of a (classical) dynamical system, this movement is in general irreversible (the entropy increases). But also a morphism $F \in$ Mor $\mathscr{C}(X,S,P)$ is an irreversible motion since, although for $u,v \in U$ the relative information is constant, for other possible states on (X,S) (if P is not complete) this is not necessarily true, and even for $u,v \in U$ F^{-1} not always exists. Only any $F \in$ Mor $\mathscr{C}_o(C,X,P)$ is always reversible. We may also say that D-geometry and R-geometry, generally D_f-geometry with condition $f(1)=0$, are invariant with respect to morphisms of categories $\mathscr{C}(X,S,P)$ and $\mathscr{C}_o(X,S,P)$ in a similar sense as all Klein geometries are invariant with respect to a group, cf. [5].

More explicitly, the mentioned (classical statistical) dynamical systems can be defined, in analogy to the usual definition [21] p. 232, as a one-parameter (parameter t = time) semigroup of class $\mathscr{C}_o$, i.e., $\{T_t : t \geq 0\} \subset$ Mor $\mathscr{C}(X,S,P)$ such that

$$T_t T_s = T_{t+s} \, , \; t,s \geq 0 \, , \tag{44}$$

$$T_o = I \text{ (identity map)} \, , \tag{45}$$

$$\text{s-}\lim_{t \to t_o} T_t = T_{t_o} \text{ (strong limit in R-topology)} \, , \; t_o \geq 0. \tag{46}$$

The difference with the usual definition is that in (X,S,P) the maps are not necessarily linear, Mor $\mathscr{C}(X,S,P) \neq L(X,X)$, in general.

Accordimg to the above definition the dynamical systems, although statistical in the sense that they concern statistical states, are in another , more abstract, Max Born's sense, deterministic (determinism not of events, but of probability distributions in the time evolution). It is possible, however, and this is our new proposition (new according to the knowledge of the author), to define also "strongly statistical" dynamical systems, or (strongly) stochastic dynamical systems, using the concept of Markov morphisms (transition probability kernels), well-known in theory of testing of statistical hypotheses, cf. [5] p. 26, 76. A Markov morphism between measurable spaces (X,S) and (Y,T) is a function $M : X \times T \longrightarrow R^1_+$ denoted by $M(x,A)$ or $M(x,dy)$, $x \in X$, $y \in Y$, $A \in T$, such that by fixed $A \in T$ M is an S-measurable function on X and by fixed $x \in X$ M is a probability measure on T :

$$\int_Y M(x,dy) = 1 , \quad x \in X. \tag{47}$$

In (11) above we have shown how a transformation F (7) on X induces a map for probability measures on S, so finally a map from P to Q. Now we define the latter by means of a Markov morphism M

$$n_u(A) = \int_X M(x,A)m_u(dx) = \int_X M(x,A)f_u(x)m(dx) , \quad A \in T, \ u \in U, \tag{48}$$

$$n(A) = \int_X M(x,A)m(dx) , \quad A \in T. \tag{49}$$

Putting in place of the class of all maps F the class of all maps M, as the class of morphisms of a category, we obtain instead of $\mathcal{C}(X,S,P)$ the category of Markov morphisms $\mathcal{M}(X,S,P)$ (the class of objects remains the same). Physically, Markov morphism can describe stochastic dynamical processes in the strong sense, i.e., one obtains a stochastic dynamical system as a one-parameter semigroup (45)-(46) with $\{T_t : t \geq 0\} \in \mathrm{Mor}\,\mathcal{M}(X,S,P)$.

The category $\mathcal{M}(X,S,P)$ generates also the corresponding geometry of stochastic evolution, but because of lack of space we cannot go here into details of this theory. But it is actually not necessary since this is only a reinterpretation of the well-known theory presented in [5]. Another reinterpretation is the theory of information channels, where, however, up to now the geometrical methods were not applied.

Here, as in [5], it is shown the way how to do it.

Here we cannot also go into details of the quantum case, cf. [12]. Roughly speaking, we have in this case to replace probability densities $f_u(x)$ by density operators ϱ_u. Formally, all goes in the same way as far we use additive operators and operations or multiplications by numbers. The problem, however, is untrivial when we have multiplication of states, as is the case, e.g., of multiplicative means (37). The proper quantum generalization is in this case as follows

$$[p,u,v] = \varrho_{uv,p} = Z^{-1}(p)\,\tau^{\frac{1}{2}}(\tilde{\varrho}_u)^p(\tilde{\varrho}_v)^{1-p}\,\tau^{\frac{1}{2}},$$

$$\tau = \varrho_u + \varrho_v \,, \quad \tilde{\varrho} = \tau^{-\frac{1}{2}}\varrho\,\tau^{-\frac{1}{2}} \,, \tag{50}$$

$$Z(p) = \mathrm{Tr}\,[\,\tau^{\frac{1}{2}}(\tilde{\varrho}_u)^p(\tilde{\varrho}_v)^{1-p}\,\tau^{\frac{1}{2}}\,] \,. \tag{51}$$

The author is greatly indebted to S.L.Woronowicz for the help in establishing the formula (50)-(51).

Remarks. The family of Csiszár's f-divergencies (27), although it defines some family of "quasi-geometries", not always has an information-theoretical interpretation, as we already mentioned before formula (31). This fact, however, is also seen in the impossibility of obtaining a family of measures of uncertainty as we mistakenly tried in formula

$$J_f(E) = -\,f(p(E)) \,, \quad I_f(E) = -\,p(E)\,f(p(E)) \,, \quad 0 \leq p(E) \leq 1 \,, \tag{52}$$

as a continuation of formula (31) and (32). Nevertheless, there exists a smaller, one-parameter family of f-functions which has this property in some generalized sense, namely, the family of the so- called structural α-entropies of Havrda-Chavrát-Daróczy [22],[23] defined by

$$f(w) = F_\alpha(w) = \frac{w(w^{\alpha-1}-1)}{1-e^{1-\alpha}} \,, \quad \alpha > 0 \,, \ \alpha \neq 1 \,, \ w \geq 0 \,, \tag{53}$$

which fulfils the condition $f(1)=0$ and for $\alpha \to 1$ goes over into the Boltzmann-Shannon function $f(w) = w \ln w$ (28).

Instead of (52) we can then write

$$J_\alpha(E) = \frac{1 - [p(E)]^{\alpha-1}}{1 - e^{1-\alpha}} \quad , \quad I_\alpha(E) = \frac{p(E)(1-(p(E))^{\alpha-1})}{1 - e^{1-\alpha}} \quad ,$$

$$0 \leqslant p(E) \leqslant 1 \ . \tag{54}$$

It is interesting that for $\alpha=2$, while (54) only unessentially differs from (32) (by a positive multiplicative constant), the corresponding $I_{Xf}(u:v) = I_2(u:v)$ (27) essentially differs from the variation distance R (24).

References

[1] Alberti,P.M. and A.Uhlmann: Dissipative motion in state space. Leipzig 1981.

[2] Aczél,J. and Z.Daróczy: On measures of information and their characterizations. New York 1974.

[3] Bar-Hillel,Y.: Language and information. New York 1964.

[4] Barra,J.R.: Notions fondamentales de statistique mathématique. Paris 1971.

[5] Chentsov,N.N.: Statistical decision rules and optimal inference. Moscow 1971.

[6] Csiszár,I.: On topological properties of f-divergencies. Studia Sci. Mat. Hung. 2(1967), 329-339.

[7] Csiszár,I.: I-divergencies geometry of probability distributions and minimization problem. Ann. Probab. 3(1975), 146-158.

[8] Fisz,M.: Probability theory and mathematical statistics. Warszawa 1967.

[9] Fisher,R.A.: On the mathematical foundations of theoretical statistics. Phil. Trans. Roy. Soc. London A 222(1922), 309-368; Contributions to mathematical statistics. New York 1950, paper 10.

[10] Ingarden,R.S. and K.Urbanik: Quantum informational thermodynamics. Acta Phys. Polon. 21(1962), 281-304.

[11] Ingarden,R.S.: Information theory and thermodynamics of light. Part II. Principles of information thermodynamics. Fortschr.

Phys. 13(1965), 755-805.

[12] Ingarden,R.S.: Information geometry in functional spaces of classical and quantum finite statistical systems. Intern. J. Engineer. Sci. 19(1981), 1609-1633.

[13] Jaynes,E.T.: Information theory and statistical mechanics. Phys. Rev. 106(1957), 620-630 , 108(1957), 171-182.

[14] Kullback,S.: Information theory and statistics. New York 1959.

[15] Lehmann,E.L.: Some principles of the theory of testing hypotheses. Ann Math. Statist. 21(1950), 1-26.

[16] Lehmann,E.L.: Testing statistical hypotheses. New York 1959.

[17] Morse,N. and R.Sackstecker: Statistical isomorphism. Ann. Math. Statist. 37(1966), 203-219.

[18] Onicescu,O.: Energia informationala. Stud. Cercet. Mat. 18(1966), 1419-1470.

[19] Sierpiński,W.: General topology. Toronto 1966.

[20] Wehrl,A.: General properties of entropy. Rev. Mod. Phys. 50(1978), 221-260.

[21] Yosida,K.: Functional analysis. Berlin 1971.

[22] Havrda,I. and F.Chavrát: Quantification method of classification processes, concept of structural a-entropy. Kybernetika (Prague) 3(1967), 30-35.

[23] Daróczy,Z.: Generalized information functions. Information and Control 16(1970), 36-51.

ON STOCHASTIC PARTIAL DIFFERENTIAL EQUATIONS OF PARABOLIC TYPE

G.Jetschke[1], W.Linde[1], R.Manthey[1]

We review different approaches to parabolic equations (related to physics) with a multiparameter "white noise" inhomogeneity, which turn out to be equivalent. Some interesting properties concerning the solutions are given.

1. Introduction.

1.1. Modern physics deals more and more with nonlinear problems amongst them the description of self-organization in natural systems is very important and must necessarily include the influence of random fluctuations [1].

We will consider the following nonlinear parabolic equation

$$\partial u(t,x)/\partial t = D\cdot(\partial^2 u/\partial x^2) + f(u) + c\cdot\xi(t,x) \quad , \qquad (1a)$$

which models the interaction of local production by chemical reactions (with the nonlinear rate function f) and spatial transport by diffusion (with diffusion constant D>0) of a substance under the conditions

$$t \in [0,T] \equiv \mathcal{T},\ x \in [0,L] \equiv \mathcal{D}\ ,\ u(t,0) = u(t,L) = 0\ ,\ u(0,x) = u_0(x). \quad (1b)$$

Internal fluctuations are modelled by adding a random source term, which is, somewhat idealized, a spatio-temporal white noise , i.e. a Gaussian random field with $E\xi_{tx}=0$, $E\xi_{tx}\xi_{t'x'} = \delta(t-t')\cdot\delta(x-x')$, the constant c controlling the strength of the noise.

The problem is to develop a mathematical calculus for treating such an (heuristic) equation. There are (at least) two different approaches:

1) $\{u_{tx}\}$ is a real number-valued field with two parameters t,x,
2) $\{u_t(.)\}$ is a function-valued process with one parameter t.

1.2. In more realistic (but also more complicated) models ξ should be substituted by a "coloured" noise ζ with $E\zeta_{tx}=0$ and (piecewise) continuous paths (f.e. a Gaussian random field with a suitable covariance $E\zeta_{tx}\zeta_{t'x'} = R(t-t',x,x')$).

Shall be in the following for the time being $f\equiv 0$ (the nonlinear case

1) Friedrich-Schiller-Universität, Sektion Mathematik, DDR-6900 Jena

is treated in section 5). Then the integration of (1) can be done pathwise,

$$u_{tx}^{(\zeta)} = \int_0^L G(t,x,y)u_o(y)dy + c\cdot\int_0^t \int_0^L G(t-s,x,y)\zeta(s,y)dyds \ , \quad (2)$$

where G is the Green's function of this problem.

2. Real-valued field approach.

2.1. The basic object is the two-parameter Wiener field $\{W_{tx}\}$, $(t,x)\in \mathcal{T}\times\mathcal{D}$, which is Gaussian with $EW_{tx}=0$ and $EW_{tx}W_{t'x'}= =(t\wedge t')\cdot(x\wedge x')$. It has a.s. continuous sample paths.
Now one can introduce a stochastic integral of non-random functions $g\in\mathcal{L}^2(\mathcal{T}\times\mathcal{D})$ with respect to W, $\int_M g_{sy}dW_{sy}$, where $M\subset\mathcal{T}\times\mathcal{D}$ is a Borel set, which can be understood as an integral with respect to the random probability measure $\nu(\{[t_1,t_2)\times[x_1,x_2)\})=W_{t_2x_2}-W_{t_s x_1}-W_{t_1x_2}+W_{t_1x_1}$.
If we choose $M=[0,t]\times[0,x]$ and vary t and x we get a random Gaussian field in (t,x).

2.2. We define: The (mild) solution of equation (1) shall be

$$u_{tx} = \int_0^L G(t,x,y)u_o(y)dy + c\cdot\int_{[0,t]\times\mathcal{D}} G(t-s.x.y)dW_{sy} \quad (3)$$

and motivate this by the fact that (at least) all finite-dimensional distributions of $\{u_{tx}^{(\zeta^{(n)})}\}$ converge to those of $\{u_{tx}\}$ if $\zeta_{tx}^{(n)}\to\zeta_{tx}$ in the sense of generalized functions ([2], [5]). The solution u_{tx} has a.s. continuous sample paths and fulfils the boundary and initial conditions (1b) ([2]).

3. Function-valued process approach.

3.1. The choice of a suitable function space is of great importance but for simplicity we choose the Hilbert space $H=\mathcal{L}^2(\mathcal{D})$, i.e. $u_t(.)\in H$, $\forall t\in\mathcal{T}$. The basic object is now the Wiener cylinder process $\{z_t\}$, $t\in\mathcal{T}$,

$$z_t = \sum_{i=1}^{\infty} b_i(t)\cdot e_i \ ,$$

where $\{e_i\}_{i=1}^{\infty}$ is an orthonormal basis in H and $\{b_i(t)\}_{i=1}^{\infty}$ are independent Wiener processes in R^1 (with variance 1).

Remark that this definition looks rather formally but there are other possibilities to introduce z_t, especially as a random linear functional in $D'(\mathcal{T}\times\mathcal{D})$ [3]. Regarded on $H\{z_t\}$ only induces a Gaussian cylinder measure with expectation 0 and covariance operator $C(z_t,z_{t'})=(t\wedge t')I$. A stochastic integral of non-random integrands F with respect to z can be defined in several ways, but here it will be sufficient to assume that $F\in\mathcal{L}^2(\mathcal{T};\mathcal{L}_{(2)}(H))$, i.e. F_t is a Hilbert-Schmidt operator $\forall t\in\mathcal{T}$ and

$$\int_0^T \|F_t\|^2_{\mathcal{L}_{(2)}(H)}\,dt<\infty,$$

and to set

$$\int_0^t F_s dz_s = \sum_{k=1}^{\infty} e_k\cdot\Big[\sum_{i=1}^{\infty}\int_0^t (F_s e_i,e_k)_H db_i(t)\Big].$$

This stochastic integral is a H-valued random process in t [3]. (Another way is to allow $F_t\in\mathcal{L}_{(2)}(K,H)$, where K is a scaled Hilbert space on which z_t is a well-defined process with covariance operator belonging to trace class [4]).

3.2. We define: The (mild) solution of equation (1) shall be

$$u_t(.) = T_t u_o + c\cdot\int_0^t T_{t-s}dz_s \quad , \tag{4}$$

where $\{T_t\}$ is the semigroup generated by $D\cdot(d^2/dx^2)$, i.e.

$$(T_t h)(x) = \int_0^L G(t,x,y)h(y)dy.$$

It turns out that $u_t(.)$ has a.s. continuous sample paths [3].

4. Equivalence of both approaches and properties of the solution.

The most important results are the following [5]:

4.1. The approaches discussed in 2. and 3. (and introduced independently in the literature) are essentially the same because $\int_{[0,t]\times\mathcal{D}} G(t-s,.,y)dW_{sy}$ and $\int_0^t T_{t-s}dz_s$, respectively, have the same distribution (on a suitable function space) and, consequently, the same holds for $u_t(.)$ and $u_{t(.)}$, resp., defined by (3) and (4). (The underlining reason is that we may put $z_t = \int_0^t \xi ds$, $\overline{w}_{tx} =$

$=\int_0^t \int_0^x \xi \, dyds$ in the sense of generalized functions, i.e. z_t is only spatially "white".)

4.2. The solution $\{u_{tx}\}$, $(t,x) \in \mathcal{T} \times \mathcal{D}$, is a Gaussian random field with

$$Eu_{tx} = \int_0^L G(t,x,y)u_o(y)dy \quad ,$$

$$C(u_{tx}, u_{t'x'}) = c^2 \cdot \int_0^{t \wedge t'} \int_0^L G(t-s,x,y)G(t'-s,x',y)dyds$$

and a.s. continuous paths or, equivalently $u_t(.)$, $t \in \mathcal{T}$, is a Gaussian process with $Eu_t = T_t u_o$ and $C(u_t, u_{t'}) = c^2 \cdot \int_0^{t \wedge t'} T_{t-s} T_{t'-s} ds$ which actually can be considered on $C(\mathcal{D})$ with a.s. continuous paths.

4.3. Now let the initial state $u_o(.)$ be random but independent of the noise $z_t(.)$. Let us decompose $u_o = (u_o - u^i) + u^i$, where u^i is Gaussian with $Eu^i = 0$ and $C(u^i) = -(c^2/2) \cdot (d^2/dx^2)^{-1}$. Then we have that $\lim E \| T_t(u_o - u^i) \|_H^2 = 0$ and $T_t u^i + c \cdot \int_0^t T_{t-s} dz_s$ is a stationary process in t, i.e. $\{u_t\}$ is asymptotically stationary.

4.4. The solution u_t is a homogeneous Markov process with transition probability $Q_{t-s}(h,B) = P(\{u_t \; B \mid u_s = h\})$, $0 \leq s \leq t$, given by the distribution of $T_{t-s}h + c \cdot \int_s^t T_{t-r} dz_r$, and the unique invariant measure is a Gaussian measure μ_G, given by the distribution of u^i in 4.3. The main reason for this fact is that u_t, knowing u_s, depends only on z_r, $s \leq r \leq t$, and that the (non-random) diffusion equation smears out any initial properties.

A short remark seems to be necessary: For two-parameter fields several definitions of a suitable Markov property are possible but they have no direct physical interpretations. For function-valued processes, however, the usual Markov property reflects the causality principle of physics.

5. Nonlinear equations.

5.1. The problem (1) becomes (in general) nonlinear if f is non-vanishing. Then, besides some measurability conditions, the solution u_t

shall fulfil the integral equation

$$u_t = T_t u_o + \int_0^t T_{t-s} f(u_s) ds + c \cdot \int_0^t T_{t-s} dz_s \quad . \tag{5}$$

5.2. If f fulfils a certain Lipschitz condition the solution of (5) exists uniquely and is a homogeneous Markov process with a.s. continuous paths [6].

5.3. If f fulfils a certain monotony condition then the solution u_t of (5) is an asymptotically stationary process, i.e. every initial distribution converges for $t \longrightarrow \infty$ to the invariant distribution P^i which is absolutely continuous with respect to the invariant (Gaussian) measure μ_G of the linear problem and

$$(dP^i/d\mu_G)(u) = \text{const} \cdot \exp \left[\frac{2}{c^2} \cdot \int_0^L V(u(x)) dx \right] ,$$

$$dV(u)/du = f(u) \quad [7] .$$

6. Remarks.

Similar results hold if $D \cdot (d^2/dx^2)$ is substituted by a more general linear operator A which generates a semigroup $\{T_t\}$. Also result 5.3. is expected to remain true if f is non-monotone.

The next step must be to get more detailed information about u_t, but the essential generalization will be the case of state-dependent strength of the noise, i.e. $c = c(u)$, which needs a more refined definition of stochastic integrals.

References.

[1] Haken,H.: Synergetics.An Introduction. Springer Series in Synergetics 1 (1978).

[2] Walsh,J.B.: A stochastic model of neural response. Adv. Appl. Prob. 13(1981), 231-281.

[3] Miyahara,Y.: Infinite dimensional Langevin equation and Fokker-Planck equation. Nagoya Math. J. 81(1981), 177-223.

[4] Arnold,L., R.F.Curtain and P.Kotelenez: Linear stochastic evolution equation models for chemical reactions. In: Stochastic Nonlinear Systems in Physics, Chemistry and Biology, ed. by L.Arnold and R.Lefever. Springer Series in Synergetics 8(1981),174-184.

[5] Jetschke,G., W.Linde and R.Manthey: forthcoming paper.

[6] Arnold,L., R.F.Curtain and P.Kotelenez: Nonlinear stochastic evolution equations in Hilbert space. Univ. Bremen, Report 17(1980).

[7] Marcus,R.: Parabolic Ito Equations with Monotone Nonlinearities. J. Funct. Anal. 29(1978), 275-286.

THE THEORY OF INVARIANT PARTITIONS FOR Z^d - ACTIONS

B. Kamiński[1)]

The theory of invariant partitions for single automorphisms has been developed by V.A.Rohlin and Y.A.Sinai in [6](see also [5]). It plays an important role in the ergodic theory. Our aim is to present an extension of the main results of the Rohlin-Sinai theory to Z^d - actions, $d \geq 1$. To avoid the technical difficulties we consider only the case d=2. It is easy to generalize the results to arbitrary Z^d - actions, $d \geq 1$. The full proofs of the theorems and examples are contained in [3] and [4].

1. Preliminaries

Let $(X, \mathcal{B}, \mu)$ be a Lebesgue space, $\mathcal{M}$ - the set of all measurable partitions of X and $\mathcal{Z}$ - the set of measurable partitions with finite entropy. If $\alpha \in \mathcal{M}$ then $L^2(\alpha)$ will denote the subspace of $L^2(X, \mu)$ consisting of all functions constant on elements of α.

Let S be an automorphism of X and $\sigma \in \mathcal{M}$ such that $S\sigma = \sigma$. For every $\alpha \in \mathcal{M}$ we put

$$\alpha_S^- = \bigvee_{n=1}^{\infty} S^{-n}\alpha \ , \quad \alpha_S = \bigvee_{n=-\infty}^{\infty} S^n\alpha \ , \quad \tau(\alpha, S) = \bigwedge_{n=0}^{\infty} S^{-n}\alpha \ ,$$

$$h_\sigma(\alpha, S) = H(\alpha \mid \alpha_S^- \vee \sigma) .$$

The σ-relative entropy of S is defined by the formula:

$$h_\sigma(S) = \sup_{\alpha \in \mathcal{Z}} h_\sigma(\alpha, S).$$

By the Pinsker closure of S with respect to σ we mean the join of all partitions $\alpha \in \mathcal{Z}$ such that $h_\sigma(\alpha, S) = 0$ and we denote it by $\bar{\sigma}$.

We have given in [2] the following properties of $\bar{\sigma}$: $\sigma \leq \bar{\sigma}$, $S\bar{\sigma} = \bar{\sigma}$, $\bar{\bar{\sigma}} = \bar{\sigma}$, if T is an automorphism commuting with S then $\overline{T\sigma} = T\bar{\sigma}$. One can also prove the following lemmas.

<u>Lemma 1.</u> There exists $\zeta \in \mathcal{M}$ with the properties:

(i) $\sigma \leq S^{-1}\zeta \leq \zeta$,

(ii) $\zeta_S = \varepsilon$,

(iii) $\tau(\zeta, S) = \bar{\sigma}$,

(iv) $h_\sigma(\zeta, S) = h_\sigma(S)$.

<u>Lemma 2.</u> For every $\alpha \in \mathcal{Z}$, $\sigma_i \in \mathcal{M}$, $S\sigma_i = \sigma_i$, i=1,2, we have

$$h_{\bar{\sigma}_1 \vee \sigma_2}(\alpha, S) = h_{\sigma_1 \vee \sigma_2}(\alpha, S).$$

Now, let G be an abelian free group of rank 2 of automorphisms

---

1) Nicholas Copernicus University, Institute of Mathematics, ul. Chopina 12/18, Toruń, Poland.

of $(X, \mathcal{B}, \mu)$. We assume in the sequel that G is aperiodic. For the definitions and the basic results of the entropy theory for finitely generated abelian groups of automorphisms the reader is referred to [1].

2. Exhaustive partitions

Let (T,S) be an ordered pair of independent generators of G. For every $\zeta \in \mathcal{M}$ we put

$$\zeta^- = \zeta_S^- \vee (\zeta_S)_T^- , \quad \zeta_G = (\zeta_S)_T .$$

The partition ζ is said to be totally invariant if $S^{-1}\zeta = \zeta$ and $T^{-1}\zeta = \zeta$.

We say that ζ is (T,S) - invariant if $S^{-1}\zeta \leqslant \zeta$ and $T^{-1}\zeta_S \leqslant \zeta$. Let us observe that ζ is (T,S) - invariant iff $T^k S^l \zeta \leqslant \zeta$ for $(k,l) \prec (0,0)$ where $\prec$ denotes the lexikographical order in Z^2.

The concept of the invariance is a natural generalization of the concept of the invariance for single automorphisms. However, for our purposes we shall need a stronger property than the invariance - namely the so called strong invariance.

The partition ζ is said to be (T,S) - strong invariant if $S^{-1}\zeta \leqslant \zeta$ and $\tau(\zeta, S) = T^{-1}\zeta_S$.

Every strong invariant partition is of course invariant. However, the converse implication is not true.

<u>Definition.</u> A partition $\zeta \in \mathcal{M}$ is (T,S) - exhaustive if it is (T,S) - strong invariant and $\zeta_G = \varepsilon$.

<u>Lemma 3.</u> If $\zeta \in \mathcal{M}$ is (T,S) - strong invariant, $\alpha \in \mathcal{Z}$, $\alpha \leqslant T^p \zeta_S$ for some natural number p and $h(\alpha, G) = 0$ then $\alpha \leqslant \tau(\zeta_S, T)$.

The proof of this lemma is contained in the proof of Théorème 5.1 in [1].

<u>Theorem 2.</u> For every (T,S) - exhaustive partition ζ we have $\tau(\zeta_S, T) \geqslant \pi(G)$.

<u>Proof.</u> Let $\alpha \in \mathcal{Z}$ and let $\alpha \leqslant \pi(G)$, i.e. $h(\alpha, G) = 0$. Let $\beta \in \mathcal{Z}$ and $\beta \leqslant T^p \zeta_S$ where p is a certain natural number.

We denote by G_k the group generated by the automorphisms T^k and S^k, where k is an arbitrary natural number.

Using the generalized Pinsker formula (see formula (19) in [1]) for α, β, G_k and the equality $h(\alpha, G) = 0$ we have

$$h(\beta, G_k) = H(\beta \mid \beta_{G_k}^- \vee \alpha_{G_k}), \quad k \geqslant 1. \tag{1}$$

Hence we obtain

$$H(\beta \mid \alpha) \geqslant H(\beta \mid \beta_{G_k}^- \vee \alpha_{G_k}) = h(\beta, G_k) \geqslant H(\beta \mid S^{-k+1}\beta_S^- \vee T^{-k+1}(\beta_S)_T^-), \tag{2}$$

$k \geqslant 1$. If $k \to \infty$ then from (2) follows

$$H(\beta \mid \alpha) \geqslant H(\beta \mid \hat{\beta}), \tag{3}$$

where $\hat{\beta} = \bigwedge_{k=0}^{\infty} (S^{-k}\beta_S^- \vee T^{-k}(\beta_S)_T^-)$.

Now our purpose is to prove

$$\hat{\beta} \leqslant \tau(\zeta_S, T). \tag{4}$$

Let $\gamma \in \mathcal{Z}$ and $\gamma \leqslant \hat{\beta}$. Conze in [1] has shown that $\hat{\beta} \leqslant \pi(G)$. Hence $\gamma \leqslant \pi(G)$, i.e. $h(\gamma, G) = 0$. Since $\beta \leqslant T^p\zeta_S$ and ζ is (T,S) - invariant we have $\hat{\beta} \leqslant T^p\zeta_S$ and so $\gamma \leqslant T^p\zeta_S$. It follows from lemma 3 that $\gamma \leqslant \tau(\zeta_S, T)$ and so (4) is satisfied. From (3) and (4) follows

$$H(\beta | \alpha) \geqslant H(\beta | \tau(\zeta_S, T)). \tag{5}$$

Since $\zeta_G = \varepsilon$ we see that (5) is satisfied for β belonging to a dense subset of $\mathcal{Z}$. Hence it is satisfied for all $\beta \in \mathcal{Z}$. Taking $\beta = \alpha$ we obtain the desired result. Thus we have proved that $\tau(\zeta_S, T) \geqslant \pi(G)$ what completes the proof.

3. Perfect partitions

Theorem 3. For every pair (T,S) of independent generators of G there exists $\eta \in \mathcal{M}$ with the properties

(a) $S^{-1}\eta \leqslant \eta$, $T^{-1}\eta_S \leqslant \eta$,

(b) $\eta_G = \varepsilon$,

(c) $\tau(\overline{\eta_S}, T) \leqslant \pi(G)$,

(d) $h(G) = H(\eta | \eta^-) = H(\eta | S^{-1}\eta)$.

Proof. Let $\alpha_k \in \mathcal{Z}$, $k \geqslant 1$ be such that $\alpha_k \nearrow \varepsilon$. Using the property 8 of the mean entropy ([1]) one can construct, similarly as in the proof of theorem 12.5 in [6], a sequence $n_1 < n_2 < \ldots$ of natural numbers such that for

$$\zeta_p = \bigvee_{k=1}^{p} T^{-n_k}\alpha_k$$

the inequality

$$H(\zeta_p | \zeta_p^-) - H(\zeta_p | \zeta_{p+s+1}^-) < \frac{1}{p}, \quad s \geqslant 0 \tag{6}$$

holds.

We put $\eta = \zeta \vee \zeta^-$ where $\zeta = \bigvee_{p=1}^{\infty} \zeta_p$. It is easy to check that η satisfies (a) and (b).

Now we want to show that (c) holds. Since $\zeta_p^- \nearrow \zeta^- = \eta^-$ then taking in (6) the limit when $s \to \infty$ we obtain

$$H(\zeta_p | \zeta_p^-) - H(\zeta_p | \eta^-) < \frac{1}{p}, \quad p \geqslant 1. \tag{7}$$

Now, let $\alpha \in \mathcal{Z}$ and $\alpha \leqslant \tau(\overline{\eta_S}, T)$. Since the last partition is totally invariant we have $\alpha_G \leqslant \tau(\overline{\eta_S}, T)$. Hence $\alpha_G \leqslant T^{-n}\eta_S$, $n \geqslant 1$ and so $\alpha_G \leqslant (\eta_S)_T^-$. It follows from the generalized Pinsker formula that

$$h(\alpha, G) = h(\zeta_p, G) - H(\zeta_p | \zeta_p^- \vee \alpha_G) + H(\alpha | \alpha^- \vee (\zeta_p)_G). \tag{8}$$

From lemma 2 follows

$$H(\zeta_p|\zeta_p^- \vee \alpha_G) \geq H(\zeta_p|\zeta_p^- \vee \overline{(\eta_S)_T^-}) = H(\zeta_p|(\zeta_p)_S^- \vee [(\zeta_p)_S]_T^- \vee (\eta_S)_T^-) =$$

$$H(\zeta_p|(\zeta_p)_S^- \vee [(\zeta_p)_S]_T^- \vee (\eta_S)_T^-) = H(\zeta_p|\zeta_p^- \vee (\eta_S)_T^-) \geq$$

$$H(\zeta_p|\zeta_p^- \vee \eta^-) = H(\zeta_p|\eta^-)$$

and so (8) implies

$$h(\alpha, G) \leq h(\zeta_p, G) - H(\zeta_p|\eta^-) + H(\alpha|\alpha^- \vee (\zeta_p)_G) \leq \frac{1}{p} + H(\alpha|\alpha^- \vee (\zeta_p)_G). \quad (9)$$

Since $(\zeta_p)_G \nearrow \varepsilon$ then taking in (9) the limit when $p \to \infty$ we obtain $h(\alpha, G) = 0$, i.e. $\alpha \leq \pi(G)$. The property (b) may be proved similarly as in [5].

Definition. The partition $\zeta \in \mathcal{M}$ is said to be (T,S) - perfect if

(i) $S^{-1}\zeta \leq \zeta$, $T^{-1}\zeta_S \leq \zeta$,

(ii) $\zeta_G = \varepsilon$,

(iii) $\tau(\zeta, S) = T^{-1}\zeta_S$,

(iv) $\tau(\zeta_S, T) = \pi(G)$,

(v) $h(G) = H(\zeta|\zeta^-) = H(\zeta|S^{-1}\zeta)$.

Theorem 4. For every ordered pair (T,S) of independent generators of G there exists a (T,S) - perfect partition.

Proof. Let $\eta \in \mathcal{M}$ be a partition satisfying properties (a) - (d) of theorem 3. From (a) follows $T^{-1}\overline{\eta_S} \leq \overline{\eta_S}$. Now lemma 1 implies that there exists $\zeta \in \mathcal{M}$ with the properties

$$T^{-1}\eta_S \leq \zeta \leq \eta_S, \quad (10)$$

$$S^{-1}\zeta \leq \zeta, \quad (11)$$

$$\zeta_S = \overline{\eta_S}, \quad (12)$$

$$\tau(\zeta, S) = T^{-1}\overline{\eta_S}, \quad (13)$$

$$h_\sigma(\zeta, S) = h_\sigma(S), \quad \sigma = T^{-1}\overline{\eta_S}. \quad (14)$$

From (10), (12) and (b) easily follow (i) and (ii). The equalities (12) and (13) immediately imply (iii). Thus we have checked that ζ is (T,S)-exhaustive and so theorem 2 implies $\tau(\zeta_S, T) \geq \pi(G)$. It follows from (c) that $\tau(\zeta_S, T) = \tau(\overline{\eta_S}, T) \leq \pi(G)$ and so we obtain (iv). It remains to verify (v). Let $\alpha_k \in Z$ and $\alpha_k \nearrow \zeta$. Hence $\alpha_k^- \nearrow \zeta^- = S^{-1}\zeta$ and so

$$H(\alpha_k|S^{-1}\zeta) \leq H(\alpha_k|\alpha_k^-) = h(\alpha_k, G) \leq h(G), \quad k \geq 1.$$

Taking the limit when $k \to \infty$ we have $H(\zeta|S^{-1}\zeta) \geq h(G)$. Now, we may (14) write in the form

$$H(\zeta|\zeta_S^- \vee T^{-1}\overline{\eta_S}) = \sup_{\alpha \in Z} H(\alpha|\alpha_S^- \vee T^{-1}\overline{\eta_S}). \quad (15)$$

Using (10) we have $T^{-1}\overline{\eta_S} \leq \zeta_S^- \leq \overline{\eta_S}$ and then

$$H(\zeta|\zeta_S^- \vee T^{-1}\overline{\eta_S}) = H(\zeta|\zeta_S^-) = H(\zeta|S^{-1}\zeta). \qquad (16)$$

It follows from lemma 2 that

$$H(\alpha|\alpha_S^- \vee T^{-1}\overline{\eta_S}) = H(\alpha|\alpha_S^- \vee T^{-1}\eta_S)$$

and so

$$\sup_{\alpha\in Z} H(\alpha|\alpha_S^- \vee T^{-1}\overline{\eta_S}) \geq \sup_{\substack{\alpha\in Z\\ \alpha\leq\eta}} H(\alpha|\alpha_S^- \vee T^{-1}\eta_S)$$

$$\sup_{\substack{\alpha\in Z\\ \alpha\leq\eta}} H(\alpha|\eta_S^- \vee T^{-1}\eta_S) = \sup_{\substack{\alpha\in Z\\ \alpha\leq\eta}} H(\alpha|S^{-1}\eta) = H(\eta|S^{-1}\eta).$$

Comparing the above result with (d), (15) and (16) we have

$$h(G) = H(\eta|S^{-1}\eta) \leq H(\zeta|S^{-1}\zeta)$$

what completes the proof.

4. Applications

Using the theorems 2 and 4 one can prove the following

Theorem 5. The following conditions are equivalent:

(a) $h(G) = 0$,

(b) for every ordered pair (T,S) of independent generators of G the only (T,S) - exhaustive partition is ε,

(c) for every ordered pair (T,S) of independent generators of G each (T,S) - strong invariant partition is totally invariant.

Definition. The group G is said to be a Kolmogoroff group, or shortly, K-group if there exists an ordered pair (T,S) of independent generators of G and $\zeta\in\mathcal{M}$ with

(i) $S^{-1}\zeta \leq \zeta$, $T^{-1}\zeta_S \leq \zeta$,

(ii) $\zeta_G = \varepsilon$,

(iii) $\tau(\zeta, S) = T^{-1}\zeta_S$,

(iv) $\tau(\zeta_S, T) = \nu$.

It is easy to check that the two-dimensional Bernoullian group is a K-group.

The theorems 2 and 3 enable to prove

Theorem 6. G is a K-group iff it has completely positive entropy.

Remark. The necessity in the above theorem has been proved also by J.P.Conze in [1].

Let (T,S) be an ordered pair of independent generators of G. For every $\zeta\in\mathcal{M}$ and $n \geq 1$ we denote by $\mathcal{B}_n(\zeta)$ the σ-algebra generated by the partition $S^{-n}\zeta_S^- \vee T^{-n}(\zeta_S)_T^-$.

Definition. We say that G is K-mixing if for every pair (T,S) of independent generators of G, every finite partition ζ of X and every $B \in \mathcal{B}$ the equality

$$\lim_{n\to\infty} \sup_{A\in\mathcal{B}_n(\zeta)} |\mu(A\cap B) - \mu(A)\cdot\mu(B)| = 0$$

holds.

Theorem 6 and the Doob martingale convergence theorem imply

Theorem 7. G is a K-group iff G is K-mixing.

One can define in a natural way the concept of a r-fold mixing and prove that every K-group is r-fold mixing, $r \geq 1$.

Theorem 4 also implies the following

Theorem 8. If $h(G) > 0$ then G has in the subspace $L^2(X,\mu) \ominus L^2(\pi(G))$ a countable Lebesgue spectrum.

From the above theorem we obtain at once

Corollary 1. If G is a K-group then it has a countable Lebesgue spectrum.

Remark. Another proof of this fact there is in [2].

Theorem 8 enables us also to find the entropy of certain classes of dynamical systems.

Corollary 2. If G has a singular spectrum or a spectrum with finite multiplicity then $h(G) = 0$. In particular every group G with a discrete spectrum has zero entropy.

References

[1] Conze,J.P.: Entropie d'un groupe abélien de transformations. Z.Wahr.verw.Geb. 25 (1972), 11-30.

[2] Kamiński,B., Sąsiada,E.: Spectrum of abelian groups of transformations with completely positive entropy. Bull.Acad.Polon.Sci., Ser.Sci.Math.Astronom.Phys. 24 (1976), 683-689.

[3] Kamiński,B.: The theory of invariant partitions for Z^d - actions, forthcoming.

[4] Kamiński,B.: Mixing properties of two-dimensional dynamical systems with completely positive entropy, forthcoming.

[5] Rohlin,V.A.: Lectures on the entropy theory of transformations with invariant measure. Uspehi Mat. Nauk 22/5 (1967), 137, 3-56.

[6] Rohlin,V.A., Sinai,Y.G.: Construction and properties of invariant measurable partitions. Dokl.Akad.Nauk 141 (1961), 1038-1041.

STOCHASTIC STABILITY OF SOME ONE-DIMENSIONAL DYNAMICAL SYSTEMS

Gerhard Keller +)

0. It is well-known that each measure-theoretic dynamical system (Ω,B,μ,T), where $T:\Omega \to \Omega$ is a B-measurable transformation and μ is a T-invariant probability on (Ω,B), determines a stationary stochastic process $(X_n(\omega))_{n\geq 0}$ by setting $X_n(\omega):= T^n\omega$. In |2| and |3| it has been shown that some nonparametric statistical decision procedures can be applied to samples from the process (X_n) provided that the underlying dynamical system possesses a "nice" weak-Bernoulli-generator.
As in practice the observed system is subject to small perturbations the goodness of such decisions depends on how close the marginal distributions of the perturbed process are to those of the unperturbed (X_n), i.e. how stable the marginal distributions of the sample-process are under perturbations of the dynamical system. In this note I shall prove a general stability theorem and give some examples.

1. Let (Ω,B,m) be a probability space and $T:\Omega \to \Omega$ a nonsingular, measurable transformation. Suppose that at time t the state of the system is given by a probability density f ($f\geq 0$, $\int f dm = 1$). Then, at time (t+1), i.e. after one application of T, the state of the system can be described by the density $Pf:= \frac{d}{dm}(T(f\cdot m))$. P extends to a positive, linear contraction on L^1_m which preserves the integral and is called the Perron-Frobenius-operator (PFO) associated with T and m.
The positive fixed-points of P with integral 1 are just the invariant densities (with respect to m) for T. Stochastic or deterministic perturbations of the system (Ω,m,T) manifest themselves in perturbations of the operator P. The most obvious measure for the deviation of two operators P and $\tilde{P}$ is perhaps $\|P-\tilde{P}\|_1$. It turns out, however, that in the context of dynamical systems this metric is too unflexible. Therefore in |10| the following measure of deviation has been proposed:

Let V be a dense, linear subspace of L^1_m which can be made into a Banach-space by a norm $\|.\|_V$. For later purposes we also require that $\{f\in V \mid \|f\|_V \leq 1\}$ is compact in L^1_m. (One may think of the spaces of Hölder-continuous functions, functions of bounded variation, etc.). W.l.o.g. we can assume $\|f\|_1 \leq \|f\|_V$. For bounded, linear operators Q on L^1_m we define the norm $|||Q|||:=\sup\{\|Qf\|_1 \mid f\in V, \|f\|_V \leq 1\}$. Then $|||P-\tilde{P}|||$ is a suitable measure of deviation.

+) Institut für angewandte Mathematik, Universität Heidelberg, Im Neuenheimer Feld 294, D-6900 Heidelberg, BRD

We are interested in the following class of operators:

DEFINITION: Let C be the class of positive, linear contractions on L^1_m satisfying

$\int Pf\,dm = \int f\,dm \quad (f \in L^1_m)$, (1)

$P(V) \subseteq V$, and (2)

P is quasicompact and mixing as operator on V, i.e. (3)

$P = \Phi + Q$, where $\Phi Q = Q\Phi = 0$, $\Phi = \lim_{n\to\infty} P^n$ is a one-dimensional projection onto the fixed-space of P (which is contained in V), and Q maps V into V with $\|Q^n\|_V \le M \cdot q^n$ $(n \in \mathbb{N})$ for some $M>0$, $0<q<1$.

For each $S \in \mathbb{R}$ set

$V(S) := \{f \in V \mid \|f\|_V \le S,\ \|f\|_1 \le 1\}$ and

$C(S) := \{P \in C \mid P(V(S)) \subseteq V(S),\ \Phi(\{f \in L^1_m \mid \|f\|_1 \le 1\}) \subseteq V(S)\}$.

(It is easily checked that $C = \bigcup_{k\in\mathbb{N}} \bigcup_{S\in\mathbb{R}} \{P \mid P^k \in C(S)\}$.)

Examples of such operators are PFO's of certain piecewise monotonic transformations on the interval (cf. |5|, |6|, |9|) and PFO's of some piecewise expanding maps on bounded regions of the plane such as the complex β-transformation (cf. |8|).

Our main result is

THEOREM: Let $P, P_n \in C(S)$, $|||P - P_n||| \le \delta$ $(n \in \mathbb{N})$. For $l>k$ set $\Pi^l_k := P_l \circ P_{l-1} \circ \dots \circ P_{k+1}$. Then

a) $\sup_{n\in\mathbb{N}} |||\Pi^n_0 - P^n||| \le \text{const.} \cdot S \cdot \delta \cdot (1 + |\ln \delta|)$, where const. depends on P.

b) If $Ph = h$, $\int h\,dm = 1$ (i.e. $h = \Phi(1)$), then
$\sup_{n\in\mathbb{N}} \|\Pi^n_0 h - h\|_1 \le \text{const.} \cdot S \cdot \delta \cdot (1 + |\ln \delta|)$, where const. depends on P.

COROLLARY: If $P_i^{k_i} = \Phi_i + Q_i \in C(S)$, $k_i \in \mathbb{N}$ $(i=1,2)$, and $|||P_1 - P_2||| \le \delta$, then $|||\Phi_1 - \Phi_2||| \le \text{const.} \cdot S \cdot \delta\ (1 + |\ln \delta|)$, where const. depends on P_1.

REMARK: In the corollary, $|||\Phi_1 - \Phi_2|||$ can be replaced by $\|\Phi_1 - \Phi_2\|_1$. This and some more detailed results for the case where P_1 is not mixing can be found in |10|.

PROOF OF THE COROLLARY: Set $k = k_1 \cdot k_2$. Then $P_i^k \in C(S)$ $(i=1,2)$ and

$$|||\Phi_1 - \Phi_2||| \le \liminf_{n\to\infty} |||(P_1^n - Q_1^n) - (P_2^n - Q_2^n)|||$$
$$\le \sup_{n\in\mathbb{N}} |||P_1^{kn} - P_2^{kn}||| + \limsup_{n\to\infty} (|||Q_1^n||| + |||Q_2^n|||)$$
$$\le \text{const.} \cdot S \cdot \delta \cdot (1 + |\ln \delta|)$$

by the theorem and observing that $|||Q_i^n||| \le \|Q_i^n\|_V$. ∎

PROOF OF THE THEOREM:

I) Set $g_n := \sup\{ \|(\Pi_k^{k+n} - P^n)f\|_1 \mid k\in\mathbb{N}_o, f\in V(S)\}$. Then $g_n \le n\cdot S\cdot\delta$.
For $f\in V(S)$ we have

$$\|(\Pi_k^{k+m+n} - P^{m+n})f\|_1 \le \|P^m(\Pi_k^{k+n} - P^n)f\|_1 + \|(\Pi_{k+n}^{k+n+m} - P^m)\Pi_k^{k+n}f\|_1$$
$$\le M\cdot q^m\cdot 2\cdot S + g_m \text{ . (For M and q see the definition.)}$$

Hence $g_{m+n} \le M\cdot q^n\cdot 2\cdot S + g_m$ for all n, which implies
$g_n \le M\cdot q^m\cdot 2\cdot S + m\cdot S\cdot\delta$ for all n, since $g_n \le n\cdot S\cdot\delta$.
Assume w.l.o.g. that $M\ge 1$, $\delta\le 2$. Choosing m such that $q^m \sim \frac{\delta}{2M}$ it follows that $g_n \le S\cdot\delta\cdot\left(1 + \frac{|\ln\delta| + \ln 2M}{|\ln q|}\right)$ for all n. As $h\in V(S)$, this proves b) of the theorem

II) Set $d_n := \sup\{\|(\Pi_k^{k+n} - P^n)f\|_1 \mid k\in\mathbb{N}_o, f\in V, \|f\|_V \le 1, \int f\,dm = 0\}$.
Then $d_1 \le \delta$. For $f\in V$ with $\|f\|_V \le 1$, $\int f\,dm = 0$ we have

$$\|(\Pi_k^{k+n+1} - P^{n+1})f\|_1 \le \|P_{k+n+1}(\Pi_k^{k+n} - P^n)f\|_1 + \|(P_{k+n+1} - P)P^nf\|_1$$
$$\le d_n + d_1\cdot M\cdot q^n ,$$

such that $d_{n+1} \le d_n + \delta\cdot M\cdot q^n$ and hence $d_n \le \delta\cdot\left(1 + \frac{M\cdot q}{1-q}\right)$.

III) An arbitrary $f\in V$ can be decomposed as follows: $f = (f - E\cdot h) + E\cdot h$, where $h = \Phi(1)$, $E = \int f\,dm$, $|E| \le \|f\|_1 \le \|f\|_V$. We have

$\|(\Pi_o^n - P^n)(E\cdot h)\|_1 = |E|\cdot\|\Pi_o^n h - h\|_1 \le \|f\|_V\cdot g_n$, and:

$\|(\Pi_o^n - P^n)(f - E\cdot h)\|_1 \le d_n\cdot(\|f\|_V + |E|\cdot\|h\|_V) \le \|f\|_V\cdot(1+S)\cdot d_n$.

Putting both estimates together we obtain
$\|(\Pi_o^n - P^n)f\|_1 \le \|f\|_V\cdot(g_n + (1+S)\cdot d_n)$ which, together with I) and II) proves a) of the theorem.

2. In the remaining part of this note we shall see, how the hypothesis of the theorem can be checked for some piecewise expanding transformations of the interval.

We call $T:[0,1]\to[0,1]$ a Lasota-Yorke-type transformation (LYT), if there exists a partition $0 = a_o < a_1 < \dots < a_N = 1$ such that (4)
$T|(a_{i-1},a_i)$ is monotone and has a C^1-extension to $[a_{i-1},a_i]$ for $i = 1,\dots,n$,
$\inf|T'| =: \alpha > 1$, (5)
$\frac{1}{T'}$ is Lipschitz-continuous with Lipschitz-constant $\mathrm{lip}(\frac{1}{T'})$. (6)

P is the PFO associated with T and m where m denotes the Lebesgue-measure on $[0,1]$. Let $V\subseteq L_m^1$ be the space of m-equivalence classes of functions of bounded variation on $[0,1]$ with values in $\mathbb{C}$, and for $f\in V$ set $\|f\|_V := \|f\|_1 + \inf\{\text{variation}(\tilde f) \mid \tilde f = f \text{ m-a.e.}\}$. Then the assumptions of part 1 of this note are satisfied (cf. |5|). Furthermore

$$\|Pf\|_V \leq 2\cdot\alpha^{-1}\cdot\|f\|_V + C(T)\cdot\|f\|_1, \tag{7}$$

where $C(T) \leq \mathrm{lip}(\frac{1}{T'}) + \frac{2}{\alpha\cdot d_o}$, $d_o := \min\{(a_i - a_{i-1}) \mid i=1,\dots,N\}$ (cf. |12|, |11|, |13|). In many cases one can find sharper bounds for $C(T)$ avoiding the influence of a very small d_o (cf. |10|). If $\alpha > 2$, then

- T has an invariant density h, (8)
- if $(T,h\cdot m)$ is weakly mixing, then $P = \Phi + Q$ as in the definition of part 1 (cf. |7|, |9|), (9)
- $P \in C(C(T)\cdot(1 - 2\cdot\alpha^{-1})^{-1})$. (10)

Sufficient conditions for $(T,h\cdot m)$ being weakly mixing can be found in |1|.

EXAMPLE 1: Let T, T_n be weakly mixing LYT's and P, P_n their corresponding PFO's. We have

- $P, P_n \in C(S)$ for some $S \in \mathbb{R}$ according to (10) and (7),
- $|||P - P_n||| \leq 12\cdot d(T,T_n)$, where
  $d(T,T_n) := \inf\{\varepsilon>0 \mid$ there are $A \subseteq [0,1]$ and $\sigma: [0,1] \to [0,1]$ s.th. $m(A) \geq 1-\varepsilon$, σ is diffeomorphism, $T|_A = T_n \circ \sigma|_A$, and for all $x \in A$: $|\sigma(x) - x| \leq \varepsilon$, $|\frac{1}{\sigma'(x)} - 1| \leq \varepsilon\}$.
  For a proof see §4 in |10|.
- If $1 < \alpha \leq 2$, choose k big enough such that $\alpha^k > 2$ and replace P by P^k and P_n by products of the type $\Pi_{(n-1)k}^{nk}$.
- If $P, P_n \in C(S)$, $|||P - P_n||| \to 0$, and T is weakly mixing, then T_n is weakly mixing for all sufficiently big n.
- The corollary of part 1, applied to the situation of this example, gives a result similar to one in |11|.

EXAMPLE 2: Let $K: [0,1]^2 \to \mathbb{R}$ be a bistochastic kernel, i.e. $K \geq 0$ and $\int K(x,y)dx = \int K(x,y)dy = 1$. Consider the Markov-process described by the transition-density $p(x,y) = K(Tx,y)$, where T is a LYT. The evolution of the densities describing the state of the system is given by the operator $P_K: L^1_m \to L^1_m$, $P_K f(x) = \int Pf(u)K(u,x)du$ where P is the PFO of T. Assume that K satisfies the following monotonicity condition:

$$\int_{\{x\leq z\}} K(x,y_1)dx \geq \int_{\{x\leq z\}} K(x,y_2)dx \quad \text{for } y_1,y_2,z \in [0,1],\ y_1 \leq y_2 .$$

Then $P \in C(S)$ implies $P_K \in C(S)$ and

- $|||P - P_K||| \leq \|P\|_V \cdot \sup_{0\leq z\leq 1} b(z)$, where
  $b(z) := \iint_{B(z)} K(x,y)dxdy$ and $B(z) := \{(x,x) \mid x\leq z<y \text{ or } y\leq z<x\}$.
  In particular, if $K(x,y) = 0$ for $|x-y| \leq \varepsilon$, then $\sup_z b(z) \leq \|K\|_\infty \cdot \varepsilon^2$. (For proofs see §4 of 10 .)
- For two particular types of stochastic perturbations there are some results in |4| related to the corollary of part 1 in the situation of this example.

ACKNOWLEDGEMENT: The form of the theorem as presented in this note has been suggested to me by Mike Keane during the conference.

LITERATURE:

|1| R.Bowen: Bernoulli maps of the interval. Isr.J.Math.28(1977)161-168.

|2| M.Denker: Statistical decision procedures and ergodic theory. These proceedings.

|3| M.Denker/G.Keller: On U-statistics and v.Mises-functionals for weakly dependent processes. Preprint.

|4| P.Góra: On small stochastic perturbations of mappings on the unit interval. Preprint University of Warsaw.

|5| F. Hofbauer/G.Keller: Ergodic properties of invariant measures for piecewise monotonic transformations. To appear in Math.Zeitschr.

|6| F.Hofbauer/G.Keller: Equilibrium states for piecewise monotonic transformations. Preprint Univ. Heidelberg.

|7| C.T.Ionescu Tulcea/G.Marinescu: Théorie ergodique pour des classes d'opérations non complètement continues. Ann.Math.52(1950)140-147.

|8| G.Keller: Ergodicité et mesures invariantes pour des transformations dilatantes par morceaux d'une région bornée du plan. C.R.Acad.Sc. Paris, Série A, 289(1979)625-627.

|9| G.Keller: Un théorème de la limite centrale pour une classe de transformations monotones par morceaux. C.R.Acad.Sc.Paris, Série A, 291(1980)155-158.

|10| G.Keller: Stochastic stability in some chaotic dynamical systems. Preprint Univ.Heidelberg.

|11| Z.Kowalski: Invariant measures for piecewise monotonic transformations. Proc. 4th Winter-School on Prob., Karpacz, Poland, 1975, Springer Lect. Notes Math. 472.

|12| A.Lasota/J.A.Yorke: On the existence of invariant measures for piecewise monotonic transformations. Transact.A.M.S. 186(1973) 481-488.

|13| S.Wong: Some metric properties of piecewise monotonic mappings of the unit interval. Transact.A.M.S. 246(1978)493-500.

THEOREMS OF GOODWYN'S TYPE

Magda Komorníková[1)], Jozef Komorník[2)]

Goodwyn's comparison theorem (cf. [4],[5]) deals with topological entropy and measure theoretic entropy of regular probability measures on compact spaces. We show that analogical results can be proved under some weaker topological and measure theoretical conditions.

1. Entropy on pseudo-compact spaces

The notion of pseudo-compact space was introduced in [2]. Recall that the topological space X is pseudo-compact if every real continuous function on X is bounded.

Remember that the topological entropy was introduced in [1] by means of open coverings containing finite subcoverings. This definition does not require that the considered topological space is compact.

The following result was proved in [6].

Theorem 1. Let X be a Hausdorff normal and pseudo-compact space. Let T be a continuous transformation of X. Then the topological entropy

$$h(T) = \sup\{h_m(T) \quad : \quad m \in \mathcal{M}_T(X)\} \tag{1}$$

where $\mathcal{M}_T(X)$ is the system of regular probability measures on X.

2. Entropy of σ-finite measures

We introduce measure-theoretic entropy of σ-finite measure as a supremum of entropies of probability measures which are absolutely continuous with respect to μ.

Definition 1. Let T be a measurable transformation of a measurable space $(X, \mathcal{S})$.

(i) For any probability measure m on $(X, \mathcal{S})$ we put

$$h_m(T) = \sup\{\overline{H}_m(T, P) \quad : \quad P \in \mathcal{P}\} \tag{2}$$

where

$$\overline{H}_m(T,P) = \overline{\lim}\, \frac{1}{n} \cdot H_m\Big(\bigvee_{j=0}^{n-1} T^{-j}(P) \Big) ,$$

$\mathcal{P}$ is the system of all finite measurable partitions of X and H_m is the Shannon entropy.

1) Department of Cybernetics, Slovak Technical University, Gottwaldovo nám. 29, 819 00 Bratislava, Czechoslovakia

2) Department of Probability and Mathematical Statistics, Comenius University, Mlynská dolina, 816 31 Bratislava, Czechoslovakia

(ii) Let μ be a σ-finite measure on $(X, \mathcal{S})$. We put

$$h_\mu^*(T) = \sup \{ h_m(T) \quad : \quad m \in \mathfrak{M} , \quad m \ll \mu \} \tag{3}$$

where $\mathfrak{M}$ is the system of all probability measures on $(X, \mathcal{S})$.

Definition 2. Let $\mathcal{B}(X)$ be the σ-algebra generated by open subsets of the topological space X. We say that a σ-finite measure μ defined on $\mathcal{B}(X)$ is regular if for any $B \in \mathcal{B}(X)$

$$\mu(B) = \inf \{ \mu(U) \ : \ B \subset U , \ U \ \text{open} \} . \tag{4}$$

Proposition. The regularity of a σ-finite measure μ is equivalent to the condition

$$\forall B \in \mathcal{B}(X) \ : \quad 0 = \inf \{ \mu(U - B) \ : \ B \subset U \ , U \ \text{open} \} . \tag{5}$$

Proof. We can write

$$B = \bigcup_{n=1}^{\infty} B_n$$

where $B_n \in \mathcal{B}(X)$, $\mu(B_n) < \infty$.

For any n there exists an open set $U_n \supset B_n$ such that

$$\mu(U_n - B_n) < \frac{\varepsilon}{2^n} .$$

For

$$U = \bigcup_{n=1}^{\infty} U_n \quad \text{we have} \quad \mu(U - B) < \varepsilon .$$

Now we prove a weaker version of Goodwyn's theorem for entropy of regular σ-finite measures.

Theorem 2. Let T be a continuous transformation of a topological space X. Let μ be an invariant σ-finite measure on $\mathcal{B}(X)$. Then the inequality

$$h_\mu^*(T) \leq h(T) \tag{6}$$

holds.

Proof. Let $m \in \mathfrak{M}$, $m \ll \mu$. For any $\varepsilon > 0$ there exists $\delta > 0$ such that

$$\mu(B) < \delta \Longrightarrow m(B) < \varepsilon . \tag{7}$$

(The proof of this fact is standard.)

Let $P = \{ A_1, \dots, A_n \}$. There exists $\varepsilon' > 0$ such that for any $Q = \{ B_0, B_1, \dots, B_n \}$ the inequalities $m(A_i \triangle B_i) < \varepsilon$, $i = 1, \dots, n$, imply $H(P|Q) < 1$ (cf. [3]). Take δ according to (7). For any $i = 1, \dots, n$, we can take closed subsets $B_i \subset A_i$ such that $\mu(B_i - A_i) < \delta$. For any $j = 0, 1, \dots$ we have

$m(T^{-j}(A_i - B_i)) < \delta$,

therefore

$m(T^{-j}(A_i - B_i)) < \varepsilon$ and $H(T^{-j}(P)|T^{-j}(Q)) < 1$.

We have

$$\overline{H}_m(T,P) = \overline{\lim} \frac{1}{n} H_m\Big(\bigvee_{j=0}^{n-1} T^{-j}(P)\Big) \leq \overline{\lim} \frac{1}{n} H_m\Big(\bigvee_{j=0}^{n-1} T^{-j}(Q)\Big) +$$

$$+ \overline{\lim} \frac{1}{n} \sum_{j=0}^{n-1} H(T^{-j}(P)|T^{-j}(Q)) < \overline{H}_m(T,Q) + 1 .$$

Put

$U_i = B_o \cup B_i$, $i = 1,\dots,n$ and $R = \{U_1,\dots,U_n\}$.

Let $H, \overline{H}, h$ be topological analogies of $H_m, \overline{H}_m, h_m$ (cf. [7]). We have

$$\overline{H}_m(T,Q) \leq H(T,Q) \leq \overline{H}(T,R) + H(Q|R) = \overline{H}(T,R) + \ln 2 .$$

Hence

$$h_m(T) \leq h(T) + 1 + \ln 2 .$$

Taking T^n instead of T we get

$$h_m(T) = \frac{1}{n} h_m(T^n) \leq \frac{1}{n}[h(T^n) + 1 + \ln 2] = h(T) + \frac{1 + \ln 2}{n} .$$

Hence (6) holds.

REFERENCES.

[1] Adler,R.L., Konheim,A.G., MacAndrew,M.H.: Topological entropy. Trans.Amer.Math.Soc.114 (1965), 309 - 319.

[2] Bagley,R.W., Connel,E.H., McKnight,J.D.: On properties characterizing pseudo-compact spaces. Proc.Amer.Math.Soc.9 (1958), 500 - 506.

[3] Billingsley,P.: Ergodic Theory and Information. John Wiley and Sons. 1965.

[4] Goodman,T.N.T.: Topological sequence entropy. Proc. London Math. Soc. 3,29 (1974), 331 - 350.

[5] Goodwyn,L.: Comparing topological entropy with measure theoretic entropy. Amer.J.Math.94 (1972), 366 - 388.

[6] Komorníková,M., Komorník,J.: Regular measures and entropy on pseudo-compact spaces. Math.Slovaca 31,3 (1981), 297 - 309.

[7] Misiurewicz,M.: Topological conditional entropy. Studia math. 55 (1976), 175 - 200.

ON WEAK MIXING FOR SEMIGROUPS

R. Kühne [1])

Let Σ be a measurable measure preserving transformation on the finite measure space $[\Omega, \mathcal{A}, m]$, and let T be the isometry in the subspace $X = L^2 \ominus \{1\}$ of the Hilbert space $L^2 = L^2(\Omega, \mathcal{A}, m)$, defined by $(Tf)(\omega) = f(\Sigma\omega)$ $(\omega \in \Omega;\ f \in X)$. Σ is a weakly mixing transformation, i.e. $\lim_{n\to\infty} \frac{1}{n} \sum_{k=0}^{n-1} |m(\Sigma^{-k}A \cap B) - m(A)m(B)| = 0$ for all $A, B \in \mathcal{A}$, if and only if T is weakly mixing to zero, i.e. $\lim_{n\to\infty} \frac{1}{n} \sum_{k=0}^{n-1} |(T^k f, f)| = 0$ for all $f \in X$. Then, e.g. the well known spectral mixing theorem states that T is weakly mixing to zero if and only if T has no eigenvalues on the unit circle.

In the paper [2] H. Gundel generalizes this and related results to certain groups of isometries in a Hilbert space X. In this connection he defines a notion of weak mixing for such groups by the aid of a general limit of means as investigated in [5].

We will prove similar results for certain semigroups of contractions in X by using Banach means for the definition of the weak mixing. Thus in the main our definition coincides with the definition of weak mixing for general amenable semigroups given by H.A. Dye [1].

1. In this section we investigate some properties of Banach means. Let H be a abelian semigroup - in the additive notion - with zero element o, and let $B = B(H)$ be the Banach space of all bounded real functions on H with $\|f\| \doteq \sup_{t\in H} |f(t)|$ $(f \in B)$. Furthermore, let $(\tau_t)_{t\in H}$ be the special representation of H in the space of the bounded linear operators in B, defined by

$$(\tau_t f)(s) = f(s+t) \quad (t, s \in H;\ f \in B).$$

A real linear functional l on B is called a Banach mean on H if

$$l(\tau_t f) = l(f) \quad (t \in H,\ f \in B) \tag{1}$$

and

$$\inf_{t\in H} f(t) \leqq l(f) \leqq \sup_{t\in H} f(t) \quad (f \in B). \tag{2}$$

---

[1]) Technische Universität Dresden, Sektion Mathematik, DDR-8027 Dresden, Mommsenstr. 13

In order to formulate the following proposition we define on the set $K = \{\frac{1}{n}\sum_{i=1}^{n} \tau_{t_i}: t_i \in H;\ i = 1,2,\ldots,n;\ n \in N\}$ the relation $\leqq$ by

$$\sigma \leqq \tau \iff \tau K \subset \sigma K \qquad (\sigma, \tau \in K).$$

Evidently, K is equipped with the relation $\leqq$ a directed set. Then we investigate the net $(\sigma^* l_o)_{\sigma \in K}$, where l_o is defined by $l_o(f) = f(o)$ $(f \in B)$.

Proposition. A functional $l \in B^*$ is a Banach mean on H if and only if l is a weak-star cluster point of the net $(\sigma^* l_o)_{\sigma \in K}$.

Proof. We make use of the relation

$$\overline{\{\sigma^* l_o:\ \sigma \in K\}} = \overline{co}\{\tau_t^* l_o:\ t \in H\} =: A\ , \tag{3}$$

where by $\overline{\{\ldots\}}$ and $\overline{co}\{\ldots\}$ we mean the weak-star closure of $\{\ldots\}$ resp. the weak-star closed convex hull of $\{\ldots\}$.

(i) Let l be a Banach mean. To prove that $l \in A$ it is sufficient to prove that for every real a the inequality $(\tau_t^* l_o)(f) \leqq a$ implies that $l(f) \leqq a$ for all $t \in H$, $f \in B$ (cf. [3]). But this follows from $l(f) \leqq \sup_{t \in H} f(t) = \sup_{t \in H} (\tau_t^* l_o)(f)$.

Let $\mu \in K$ and let $U = U(f_1,\ldots,f_n;\ \varepsilon) = \{r \in B^*: |(r-l)(f_i)| < \varepsilon;\ i = 1,\ldots,n\}$ be a weak-star neighborhood of l. Since $l \in A$, the neighborhood $V = U(\mu f_1,\ldots,\mu f_n;\ \varepsilon)$ of l contains an element $\sigma^* l_o$ with $\sigma \in K$. By equality (1) we have $\mu^* l = l$, and for $\tau = \sigma\mu$ we obtain $\tau^* l_o \in U$ and $\tau \geqq \mu$; therefore l is a weak-star cluster point of the net $(\sigma^* l_o)_{\sigma \in K}$.

(ii) Let l be a weak-star cluster point of the net $(\sigma^* l_o)_{\sigma \in K}$. Evidently, the inequality (2) holds. We prove (1). Let $t \in H$, $\tau = \tau_t$ and $f \in B$. By the properties of l, for every $n \in N$ there exists $\sigma \in K$ such that for $\mu_n = (\frac{1}{n}\sum_{i=1}^{n} \tau^i)$ we get

$$|(\mu_n^* l_o - l)(f)| < \frac{1}{n}\ ,\quad |(\mu_n^* l_o - l)(\tau f)| < \frac{1}{n}\ .$$

Then it follows

$$|(\tau^* l - l)(f)| \leqq |(\tau^* l - \tau^* \mu_n^* l_o)(f)| + |(\tau^* \mu_n^* l_o - \mu_n^* l_o)(f)| + \\ + |(\mu_n^* l_o - l)(f)| \leqq \frac{2}{n}(1 + \| l_o \| \| f \|),$$

and therefore we obtain the validity of (1).

Remark. By the Alaoglu theorem, the proposition implies the

existence of a Banach mean on H.

2. In the following we assume that the semigroup H is contained in a locally compact abelian group G with the property

$$H \cup (-H) = G \ . \tag{4}$$

Furthermore, let X be a Hilbert space with the inner product $(.,.)$ and let $(T_t)_{t \in H}$ be a weakly continuous semigroup of linear contractions in X with $T_{t+s} = T_t T_s$ $(t,s \in H)$. Moreover, we assume that

(D) For every $x \in X$ there exists a positive definite continuous function f_x on G such that

$$f_x(t) = (T_t x, x) \quad (t \in H) \ . \tag{5}$$

For example, the condition (D) holds in the cases $H = Z^+$ (nonnegative integers), R^+ (nonnegative reals), and, for general H, if $(T_t)_{t \in H}$ is a semigroup of isometries (on connections with the existence of a unitary dilation of $(T_t)_{t \in H}$ cf. [6]).

By Bochner's theorem (cf. [5]) from (D) it follows, for every $x \in X$, that there exists a nonnegative finite regular measure μ_x on the dual group Γ [1]) of G such that

$$(T_t x, x) = \int_\Gamma \chi(t) \mu_x(d\chi) \quad (t \in H, \ x \in X) \ . \tag{6}$$

Consequently, we have

$$|(T_t x, x)|^2 = \int_\Gamma \int_\Gamma \chi(t) \psi^{-1}(t) \mu_x(d\chi) \mu_x(d\psi) \quad (t \in H, \ x \in X) \ . \tag{7}$$

First of all we shall prove some lemmas, and we denote by Λ the set of all complex valued Banach means on H (in the sense of the natural complex valued extensions of the real Banach means on H).

<u>Lemma 1</u>. If $l \in \Lambda$ and $\chi, \psi \in \Gamma$, then

$$l(\chi^{-1}(.)\psi(.)) = \begin{cases} 1 & \text{if} \quad \chi = \psi \\ 0 & \phantom{\text{if}} \quad \chi \neq \psi \end{cases} \ .$$

<u>Proof</u>. By (1), for $t \in H$ we have

$$l(\chi^{-1}(.)\psi(.)) = l(\chi^{-1}(t)\psi(t)\chi^{-1}(.)\psi(.)) = \\ = \chi^{-1}(t)\psi(t) l(\chi^{-1}(.)\psi(.)) \ .$$

If $\chi \neq \psi$, then by (4) there exists $t_o \in H$ with $\chi^{-1}(t_o)\psi(t_o) = 1$, and therefore $l(\chi^{-1}(.)\psi(.)) = 0$. If $\chi = \psi$ it follows by (2) that $l(\chi^{-1}(.)\psi(.)) = 1$.

---

[1]) with multiplication as group operation.

<u>Lemma 2</u>. If $x \in X$, $\chi \in \Gamma$ and $l \in \Lambda$, then

(i) $l(\chi^{-1}(.)(T_. x,x)) = \int_\Gamma l(\chi^{-1}(.)\psi(.))\mu_x(d\psi) = \mu_x(\{\chi\})$,

(ii) $l(|(T_. x,x)|^2) = \int_\Gamma\int_\Gamma l(\chi^{-1}(.)\psi(.))\mu_x(d\chi)\mu_x(d\psi) =$

$= \sum_{\chi\in\Gamma} \mu_x(\{\chi\})^2$.

<u>Proof</u>. We only prove the equality (i); the proof of (ii) is analogous to the proof of (i).

Let $\varepsilon > 0$ and let U be an open neighborhood of χ with

$$\mu_x(U \setminus \{\chi\}) < \varepsilon \ . \tag{8}$$

Furthermore, let C be a compact subset of $\Gamma \setminus U$ with

$$\mu_x([\Gamma \setminus U]\setminus C) < \varepsilon \ . \tag{9}$$

By the properties of Banach means proved in section 2, for every $\varphi \in C$ there exists $\sigma_\varphi \in K$ such that $|\sigma_\varphi(\chi^{-1}\varphi)(o) - l(\chi^{-1}\varphi(.))| =$ $= |\sigma_\varphi(\chi^{-1}\varphi)(o)| < \varepsilon$. By continuity arguments, there exists a neighborhood V_φ of φ with

$$|\sigma_\varphi(\chi^{-1}\psi)(o)| < \varepsilon \quad \text{for all } \psi \in V_\varphi \ . \tag{10}$$

Suppose $\sigma \geqq \sigma_\varphi$, i.e. $\sigma = \frac{1}{m}\sum_{q=1}^{m} \tau_{s_q}\sigma_\varphi$. Since $\sigma_\varphi \in K$, we have $\sigma_\varphi = \frac{1}{n}\sum_{r=1}^{n} \tau_{t_r}$ and therefore $\sigma(\chi^{-1}\psi)(o) =$ $= \frac{1}{m}\sum_{q=1}^{m} \chi^{-1}\psi(s_q)\left[\frac{1}{n}\sum_{r=1}^{n} \chi^{-1}\psi(t_r)\right]$. By (10) we obtain

$$|\sigma(\chi^{-1}\psi)(o)| \leqq |\sigma_\varphi(\chi^{-1}\psi)(o)| < \varepsilon \quad \text{for all } \psi\in V_\varphi,\ \sigma \geqq \sigma_\varphi \ .$$

The compactness of C implies $C \subset V_{\varphi_1} \cup V_{\varphi_2} \cup \ldots \cup V_{\varphi_p}$ and with $\sigma' = \sigma_{\varphi_1}\sigma_{\varphi_2}\cdots\sigma_{\varphi_p}$ we get

$$|\sigma(\chi^{-1}\psi)(o)| < \varepsilon \quad \text{for all } \psi \in C,\ \sigma \geqq \sigma' \ . \tag{11}$$

(8), (9) and (11) imply for all $\sigma \geqq \sigma'$

$$\left|\int_\Gamma [\sigma(\chi^{-1}\psi)(o) - l(\chi^{-1}\psi(.))]\mu_x(d\psi)\right| \leqq$$

$$\leqq \left|\int_C \sigma(\chi^{-1}\psi)(o)\mu_x(d\psi)\right| + \left|\int_{(\Gamma\setminus U)\setminus C} \sigma(\chi^{-1}\psi)(o)\mu_x(d\psi)\right| +$$

$$+\left|\int_{U\setminus\{\chi\}} \sigma(\chi^{-1}\psi)(o)\mu_x(d\psi)\right| + \left|\int_{\{\chi\}} [\sigma(\chi^{-1}\psi)(o)-1]\mu_x(d\psi)\right| <$$

$$< \varepsilon(2 + \mu_x(C)) \ .$$

By the proposition there exists a subnet $(\sigma_j)_{j\in J}$ of the net $(\sigma)_{\sigma\in K}$ such that $l(\int_\Gamma \chi^{-1}\psi(.)\mu_x(d\psi)) = \lim_{j\in J} \sigma_j \int_\Gamma \chi^{-1}\psi(o)\mu_x(d\psi)$.

Therefore we obtain $l(\int_\Gamma \chi^{-1}\psi(.)\mu_x(d\psi)) = \lim_{j\in J} \int_\Gamma \sigma_j(\chi^{-1}\psi)(o)\mu_x(d\psi) =$ $= \lim_{\sigma\in K} \int_\Gamma \sigma(\chi^{-1}\psi)(o)\mu_x(d\psi) = \int_\Gamma l(\chi^{-1}\psi(.))\mu_x(d\psi) = \mu_x(\{\chi\})$.

Lemma 3. Let $l \in \Lambda$ and $f \in B = B(H)$. Then the following conditions are equivalent:

(i) $l(|f(.)|) = 0$,

(ii) for each $\varepsilon > 0$ there exists a subset $A \subset H$ such that $|f(t)| < \varepsilon$ for all $t \in A$ and $l(1_A(.)) = 1$ (where 1_A denotes the indicator function of A).

Proof. (i) $\Rightarrow$ (ii): Suppose there exists an $\varepsilon > 0$ such that $l(1_A(.)) < 1$ for every $A \subset H$ with $|f(t)| < \varepsilon$ for all $t \in A$. Then for $A_\varepsilon = \{t \in H: |f(t)| \geqq \varepsilon\}$ follows $l(1_{A_\varepsilon}(.)) > 0$, and therefore

$$l(|f(.)|) \geqq l(|f(.)| 1_{A_\varepsilon}(.)) \geqq \varepsilon l(1_{A_\varepsilon}(.)) > 0 .$$

This contradicts (i).

(ii) $\Rightarrow$ (i): With ε, A and f as in (ii) we obtain

$$0 \leqq l(|f(.)|) = l(|f(.)| 1_A(.)) + l(|f(.)| 1_{H \setminus A}(.)) \leqq \varepsilon .$$

Lemma 3 implies the

Corollary. For each $l \in \Lambda$ and $f \in B$, $l(|f(.)|) = 0$ is equivalent to $l(|f(.)|^2) = 0$.

3. The semigroup $(T_t)_{t \in H}$ is called weakly mixing to zero if, for every $x \in X$, there exists a Banach mean $l_x \in \Lambda$ such that $l_x(|(T_.x,x)|) = 0$. This definition directly contains the usual definitions of weak mixing to zero in the cases $H = Z^+$, $H = R^+$. Moreover, the following results will show that this definition coincides in our case with a corresponding definition given by H.A. Dye [1] in the general case of topological amenable semigroups.

Let X_d be the subspace of X spanned by the elements $x \in X$ with the property that there exists a complex function c on H with $|c(t)| = 1$ and $T_t x = c(t)x$ for all $t \in H$, and let X_c be the orthogonal complement of X_d.

The following theorem generalizes corresponding methods and results investigated by U. Krengel [4] for the case $H = Z^+$.

Theorem. The following conditions are equivalent:

(i) $x \in X_c$,

(ii) μ_z has no atoms for all $z = T_t x$ $(t \in H)$,

(iii) $l(|(T_.z,z)|) = 0$ for all $l \in \Lambda$ and $z = T_t x$ $(t \in H)$,

(iv) $l((T_.x,y)) = 0$ for all $l \in \Lambda$, $y \in X$,

(v) for every $z = T_t x$ $(t \in H)$, there exists $l_z \in \Lambda$ with

$$l_z(|(T_.z,z)|) = 0 \ ,$$

(vi) there exists $l_x \in \Lambda$ with $l_x(|(T_.x,y)|) = 0$ for all $y \in X$.

Proof. (i)$\Rightarrow$(ii): First of all, we prove (ii) for $z = x$. Suppose, there exists $\chi \in \Gamma$ with $\mu_x(\{\chi\}) > 0$. For $\tau = \frac{1}{n}\sum_{r=1}^{n} \tau_{t_r} \in K$, we define $S_\tau = \frac{1}{n}\sum_{r=1}^{n} S_{t_r}$ with $S_t = \chi^{-1}(t)T_t$ $(t \in H)$. By ergodic theorem arguments we obtain that the net $(S_\tau x)_{\tau \in K}$ converges to an element $x_0 \in X$ with $S_t x_0 = x_0$ $(t \in H)$, and therefore $T_t x_0 = \chi(t)x_0$ $(t \in H)$. This implies $(x_0,x) = 0$, and by the proposition on Banach means it follows that

$$l[(S_.x,x)] = \lim_{\tau \in K} (S_\tau x,x) = (x_0,x) = 0 \ . \tag{12}$$

But this is a contradiction to $l[(S_.x,x)] = l[\chi^{-1}(.)(T_.x,x)] = \mu_x(\{\chi\}) > 0$ (cf. lemma 2, (i)).

Since $x \in X_c$ implies $z = T_t x \in X_c$ $(t \in H)$ (cf. e.g. [6]), the statement (ii) is proved.

(ii)$\Rightarrow$(iii): This follows from lemma 2, (ii), and the corollary of lemma 3.

(iii)$\Rightarrow$(iv): Let $l \in \Lambda$, $l(|(T_.T_t x,T_t x)|) = 0$ for all $t \in H$. The set $M = \{ y \in X\colon l(|(T_.x,y)|) = 0 \}$ is a closed linear subspace of X. Since $l(|(T_.x,T_t x)|) = l(|(T_.T_t x,T_t x)|) = 0$, we obtain $T_t x \in M$ for all $t \in H$. Then $(y,u) = 0$ $(y \in M)$ implies that $(T_t x,u) = 0$ for all $t \in H$, and therefore $u \in M$. Finally, we get $u = o$ and $X = M$.

(iv)$\Rightarrow$(v)$\Rightarrow$(iii): The first implication is evident, the second implication follows from lemma 2, (ii), and the corollary of lemma 3.

(iii)$\Rightarrow$(vi): Since (iii)$\Rightarrow$(iv), this is evident.

(vi)$\Rightarrow$(i): Suppose $x_0 \in X_d$, $T_t x_0 = c(t)x_0$, $|c(t)| = 1$ $(t \in H)$. Since $(T_t x,x_0) = (x,T_t^* x_0) = (x,c(t)^{-1}x_0) = c(t)(x,x_0)$, it results from (vi) that $|(x,x_0)| = l_x(|(T_.x,x_0)|) = 0$, hence $x \in X_c$.

In the case that $H = G$, H. Gundel [2] defines weak mixing of $(T_t)_{t \in G}$ by the aid of a certain limit of means $\lim_\alpha M_\alpha(T_t x,x)$. Then the preceding theorem and Gundel's results imply that the existence of a Banach mean l_x on G with $l_x(|(T_.x,x)|) = 0$ is equivalent to Gundel's condition $\lim_\alpha M_\alpha(|(T_t x,x)|) = 0$. Hence the weak mixing in [2] is the same as above.

Finally, from the theorem directly follows a more general version of v. Neumann's spectral mixing theorem which was proved by H.A. Dye [1] in the general case of a amenable topological semigroup H, by the aid of a refinement of the traditional Peter-Weyl theory.

Corollary (spectral mixing theorem). The semigroup $(T_t)_{t \in H}$ is weakly mixing to zero if and only if $(T_t)_{t \in H}$ has no element $x \neq 0$ with $T_t x = c(t)x$, $|c(t)| = 1$ $(t \in H)$.

References

[1] H.A. Dye, On the ergodic mixing theorem, Trans. Amer. Math. Soc. 118, 123-130 (1965).

[2] H. Gundel, Weakly and strongly mixing locally compact abelian groups od measure preserving transformations with application to abelian groups of gaussian automorphisms, Z. Wahrscheinlichkeitstheorie verw. Gebiete 49, 313-323 (1979).

[3] M. Krein and V. Šmulian, On regularly convex sets in the space conjugate to a Banach space, Annals of Math. 41, 556-583 (1940).

[4] U. Krengel, Weakly wandering vectors and weakly independent partitions, Trans. Amer. Math. Soc. 164, 199-226 (1972).

[5] W. Rudin, Fourier analysis on groups, New York - London (1962).

[6] B. Sz.-Nagy and C. Foiaş, Harmonic analysis of operators on Hilbert space, Budapest - Amsterdam - London (1970).

METRIC AND SPECTRAL ISOMORPHISMS OF MORSE DYNAMICAL SYSTEMS

J.Kwiatkowski[1]

Kakutani [1] has described a class of one-sided sequences and he discovered a detailed spectral analysis of dynamical systems induced by those sequences. A more general class of one-sided sequences, generalized Morse sequences, has been described by Keane [2]. In this lecture are considered the problems of metric and spectral isomorphisms in the class of Morse dynamical systems, i.e. the systems induced by generalized Morse sequences. First we introduce some concepts and notations used in the sequel.

Let $X = \prod_{-\infty}^{\infty} \{0,1\}$ be the set of all two-sided sequences of zeroes and ones. If $x \in X$ and $B = (b_1,\dots,b_n)$, $b_i=0,1$, is a block then $x[i,k]$, $B[i,k]$, $1 \le i \le k \le n$, will denote the blocks $(x_i,\dots,x_k)$ and $(b_i,\dots,b_k)$ respectively. Further we put $\tilde{B} = (\tilde{b}_1,\dots,\tilde{b}_n)$, where $\tilde{b}=1-b$, $b=0,1$. We will write $x[i]$ and $B[i]$ instead of $x[i,i]$ and $B[i,i]$. If $C = (c_1,\dots,c_m)$ is a block then BC means the block $(b_1,\dots,b_n,c_1,\dots,c_m)$. Now we can define a block $B \times C$ putting

$$B \times C = B^{c_1} B^{c_2} \dots B^{c_m},$$

where $B^0 = B$ and $B^1 = B$.

Let $b^0, b^1, \dots$ be finite blocks of 0's and 1's of the length at least two, starting with 0. Then we can define a one-sided sequence x by

$$x = b^0 \times b^1 \times \dots . \tag{1}$$

Let B,C be blocks, $|B| \le |C|$, and let $fr(B,C)$ be the frequency of B in C, i.e.

$$fr(B,C) = \operatorname{card}\{ j,\ 1 \le j \le |C|-|B|+1,\ C[j,j+|B|-1] = B \}.$$

Next, let $r_i = \min\left\{ \frac{1}{\lambda_i} fr(0,b^i),\ \frac{1}{\lambda_i} fr(1,b^i) \right\}$, where $\lambda_i = |b^i|$.

A sequence x defined by (1) is called a Morse sequence if $\sum_0^{\infty} r_i = \infty$

---

[1] Nicholas Copernicus University, Institute of Mathematics
ul. Chopina 12/18, Toruń, Poland.

and an infinity of the b^i is different from 0...0 and an infinity is different from 01...010 . A sequence x is called a continuous Morse sequence if either infinite many of the λ_i are even numbers or if the λ_i are odd numbers for $i \geq i_0$ and

$$\sum_0^\infty \min\{ e_0(b^i) + o_1(b^i) , o_0(b^i) + e_1(b^i)\} = \infty ,$$

where $e_k(b^i) = \frac{1}{\lambda_i} \operatorname{card}\{ j , 0 \leq j \leq \lambda_i - 1 , b^i[j] = k , j - \text{even} \}$,

$$o_k(b^i) = \frac{1}{\lambda_i} \operatorname{card}\{ j , 0 \leq j \leq \lambda_i - 1 , b^i[j] = k , j - \text{odd} \} ,$$

k=0,1. It is known [2] that if x is a continuous Morse sequence then each block B has the average relative frequency $m_x(B)$ on x, i.e.

$$m_x(B) = \lim \frac{1}{n} \operatorname{fr}(B, x[1,n]) .$$

We have $m_x(B)=m_x(\tilde{B})$ and in particular $m_x(0)=m_x(1)=\frac{1}{2}$. Obviously, m_x is an invariant measure defined by x and a Morse dynamical system (X,m_x,T) we shall denote by $\theta(x)$ (here T is the shift on X).

Now, we construct a special dynamical system, which will be isomorphic to the system $\theta(x)$. Set $Z_{\lambda_i} = \{ 0,1,\dots,\lambda_i-1 \}$, $i \geq 0$, $Z=\prod_0^\infty Z_{\lambda_i}$ and let $\mathcal{F}$ be the borelian σ-field of subsets of Z. By $\bar{p}$ we denote the product measure on $\mathcal{F}$ of $\bar{p}_0,\bar{p}_1,\dots$, where $\bar{p}_i$, $i \geq 0$, is a measure on Z_{λ_i} given by $\bar{p}_i(z_i) = \frac{1}{\lambda_i}$, $z_i \in Z_{\lambda_i}$. Let $n_i = \lambda_0 \dots \lambda_i$, $i \geq 0$ and let Λ be the group of all n_i-roots of the unity. Then Z may be identified with the dual group of Λ . Next, we mark out a subset $\bar{Z}$ of Z in the following way. If $\bar{z}=(z_0,z_1,\dots) \in Z$, then $\bar{z} \in \bar{Z}$ iff infinite many of the z_i are different from 0 and infinite many are different from λ_i-1. It is clear that $\bar{p}(\bar{Z})=1$. We define a transformation $\bar{S} : \bar{Z} \longrightarrow \bar{Z}$ by the formula

$$\bar{S}(z_0,z_1,\dots) = (0,\dots,0,z_i+1,z_{i+1},\dots) ,$$

where i is the first number such that $z_i < \lambda_i-1$. Then $(\bar{Z},\mathcal{F},\bar{p},\bar{S})$ is an ergodic dynamical system with discrete spectrum and Λ is the point spectrum of this system. It turns out [4] that the dynamical system $\theta(x)$ is metrically isomorphic to a skew product of the automorphism $\bar{S}$ with a family $S(\bar{z})$ of permutations of the two-point set $\{0,1\}$ equipped with the uniform probability. In order to describe a family $S(\bar{z})$, $\bar{z} \in \bar{Z}$, we define a function $p(\bar{z})$. Take $\bar{z}=(z_0,z_1,\ldots) \in Z$ and let $j=j(\bar{z})$ be the first number j such that $z_j < \lambda_j - 1$. Put $p(\bar{z})=0$ if $j(\bar{z})=j=0$ or if $j>0$ and the number of unities in the sequence

$$b^1[\lambda_1]\ldots b^{j-1}[\lambda_{j-1}]\, b^j[z_j+1]\, b^j[z_j+2]$$

is even. Next, we put $p(\bar{z})=1$ if $j>0$ and the above number is odd. Now we can define a family $S(\bar{z})$, $\bar{z} \in \bar{Z}$, putting $S(\bar{z})=I$ if $p(\bar{z})=0$ and $S(\bar{z})=1-I$ if $p(\bar{z})=1$. Here I means the identity on $\{0,1\}$.

We describe in the sequel a mapping f from $M = \bar{Z} \times \{0,1\}$ to X, which establishes an isomorphism between the dynamical systems $(M, S \times \{S(\bar{z})\}_{\bar{z} \in \bar{Z}})$ and $\theta(x)$. We set $j_0=z_0$, $j_i=j_i(\bar{z})=z_0+z_1n_0+\ldots+z_in_{i-1}$, $i>0$, and $k_i=n_i-j_i-1$. It is obvious that $\bar{z} \in \bar{Z}$ implies $j_{i+1} \geq j_i$, $k_{i+1} \geq k_i$, $i \geq 0$ and $j_i \longrightarrow \infty$, $k_i \longrightarrow \infty$ as $i \longrightarrow \infty$. We define a two-sided sequence $y=f(\bar{z},0)$ as follows

$$y[-j_0,k_0]=b^0,$$

and in general

$$y[-j_i,k_i] = \begin{cases} y[-j_{i-1},k_{i-1}]\,x b^i, & \text{if } b^i[z_i+1]=0 \\ \\ y[-j_{i-1},k_{j-1}]\,x b^i, & \text{if } b^i[z_i+1]=1 \end{cases} \qquad (2)$$

The sequence $y=f(\bar{z},1)$ is defined by

$$y[-j_0,k_0] = \tilde{b}^0$$

and $y[-j_i,k_i]$ as in (2).

Now, we pass on to the problems of metric and spectral isomorphisms. We shall put an additional condition on a Morse sequence x. Put

$$h_t(00,11)=fr(00,b^t) + fr(11,b^t),$$
$$h_t(01,10)=fr(01,b^t) + fr(10,b^t),$$

$t \geq 0$. We say that the sequence of blocks $\{b^0,b^1,\ldots\}$ is strong regular if there exists a positive number $\eta > 0$ such that

$$\frac{1}{\lambda_i} h_i(00,11) \geq \eta, \quad \frac{1}{\lambda_i} h_i(01,10)$$

for $i=0,1,\ldots$. Suppose that B and C are blocks of the length n_i,

$i \geqslant 0$. Define

$$d(B,C) = \frac{1}{n_i} \operatorname{card} \{ j, 1 \leqslant j \leqslant n_i, B[j] \neq C[j] \} .$$

We shall write b_i^t instead of $(b^t)^i$, where $i=0,1$.

Theorem 1. Let $x = b^0 \times b^1 \times \ldots$, $y = \beta^0 \times \beta^1 \times \ldots$ be Morse sequences such that $|b^i| = |\beta^i| = \lambda_i$, $i \geqslant 0$, and the sequences of blocks $\{ b^0, b^1, \ldots \}$ and $\{ \beta^0, \beta^1, \ldots \}$ are strong regular. Then the dynamical systems $\theta(x)$ and $\theta(y)$ are metrically isomorphic iff there exist numbers q_t, $0 \leqslant q_t \leqslant \lambda_t - 1$, and a sequence of pairs $\{ i_t, j_t \}$, $i_t, j_t = 0,1$, $t \geqslant 0$, such that

$$\sum_{t=0}^{\infty} d(\beta^t, b_{i_t}^t b_{j_t}^t [q_t+1, q_t+\lambda_t]) < \infty$$

and $$\sum_{t=0}^{\infty} \min\left(\frac{w_t}{n_t}, 1-\frac{w_t}{n_t} \right) < \infty ,$$

where $w_0 = q_0$, $w_t = q_0 + q_1 n_0 + \ldots + q_t n_{t-1}$, $t \geqslant 1$.

Remark. If a sequence of blocks $\{ b^0, b^1, \ldots \}$ satisfies a condition

$$\overline{\lim} \min \left\{ \frac{1}{\lambda_i} h_i(00,11), \frac{1}{\lambda_i} h_i(01,10) \right\} > 0 \qquad (3)$$

then there exist numbers $0 \leqslant t_0 < t_1 < \ldots$ such that the sequence of $g^0, g^1, \ldots$ is strong regular, where $g^0 = b^0 \times \ldots \times b^{t_0}$, $g^{j+1} = b^{t_j+1} \times \ldots \times b^{t_{j+1}}$, $j \geqslant 0$.

A Morse sequence x defined by blocks $b^0, b^1, \ldots$ satisfying the condition (3) is called a regular Morse sequence.

As a consequence of theorem 1 we obtain

Corollary 1. Let Morse sequences x and y satisfy the assumptions of the theorem 1 and in addition the sequence $\{ \lambda_i \}$ is bounded. Then $\theta(x)$ and $\theta(y)$ are metrically isomorphic iff $b^t = \beta^t$ for sufficiently large t.

Corollary 2. If x is a regular Morse sequence then there exists continuum Morse sequences y such that $\theta(y)$ is metrically isomorphic to $\theta(x)$ and induced measures m_y are pairwise orthogonal.

We shall occupy oneself a spectral property of Morse dynamical systems.

We begin by describing results given in [2]. Let

$$L_d = \{ g \in L^2(X,m_x) \ , \ g=\tilde{g} \} ,$$

$$L_c = \{ g \in L^2(X,m_x) \ , \ g=-\tilde{g} \} ,$$

where $\tilde{g}(y)=g(\tilde{y})$, $y \in X$ and $\tilde{y}$ is a sequence obtained with y by changing 0 to 1 and vice versa. Then $L^2(X,m_x) = L_d \oplus L_c$ and L_d, L_c are T-invariant subspaces of $L^2(X,m_x)$. Moreover, T has discrete spectrum on L_d with the eigenvalue group Λ and T has continuous spectrum on L_c . The maximal spectral type of T on L_c is determined by the function h defined by

$$h(y) = (-1)^{y_0} \ , \ y \in X .$$

In [3] a detailed description of the maximal spectral type of T on L_c is given. Now, we pass on to results containing in [5].

<u>Theorem 2.</u> The operator T has a simple spectrum on L_c.

In order to present a result concerning to the spectral isomorphism we introduce a notion of s-equivalence between the blocks of the same length.

Let B,C be blocks with $|B| = |C| = n$, $n \geq 1$, and let

$$s(k,B) = \operatorname{card} \{ i \ , \ 1 \leq i \leq n-k \ , \ B[i] \neq B[i+k] \} ,$$

$k=0,1,\ldots,n-1$. In the same way we can define the numbers $s(k,C)$. We say that the blocks B and C are s-equivalent if $s(k,B)=s(k,C)$ for $k=0,1,\ldots,n-1$. Remark that the relation of s-equivalence does not imply the identity of the blocks. For example if B = 001 and C = 011 then B is s-equivalent to C but $B \neq C$.

<u>Theorem 3.</u> Let $x = b^0 \times b^1 \times \ldots$, $y = \beta^0 \times \beta^1 \times \ldots$ be Morse sequences such that $|b^i| = |\beta^i| = \lambda_i$, $\{\lambda_i\}$ is bounded and the sequences of blocks $\{b^0, b^1, \ldots\}$, $\{\beta^0, \beta^1, \ldots\}$ are strong regular. Then the shift dynamical systems $\theta(x)$ and $\theta(y)$ are spectrally isomorphic iff the blocks b^j and β^j are s-equivalent for sufficiently large j.

References

[1] Kakutani, S.: Strictly ergodic symbolic dynamical Systems. Proc. 6th Berkeley Symp. 2 , 319-326.

[2] Keane, M.: Generalized Morse Sequences. Z. Wahrscheinlichkeitstheorie verw. Geb. 10 (1968), 335-353.

[3] Keane, M.: Strongly Mixing g-Measures. Inv. math. 16/4 (1972), 309-324.

[4] Kwiatkowski, J.: Isomorphism of some Morse dynamical systems. To appear.

[5] Kwiatkowski, J.: Spectral isomorphism of Morse dynamical systems. To appear.

COMPATIBILITY IN LOGICS AND PRODUCT MEASURABILITY

Tibor Neubrunn[1)]

It is shown that under certain compatibility conditions some results on measurability of functions of two variables may be transferred to special kinds of quantum logics.

A (quantum) logic is defined (see [7]) to be a partial ordered set $(L \leq)$ with the smallest element 0 and the greatest element 1. Further a complement a' is associated to any $a \in L$ such that for any $a, b \in L$

(i) $(a')' = a$,

(ii) $a \leq b$ if and only if $b' \leq a'$,

(iii) $a \vee a' = 1$, $a \wedge a' = 0$,

(iv) if $a \leq b$ then $c \in L$ exists such that $c \leq a'$ and $a \vee c = b$,

(v) if $a_n \in L$ $(n = 1,2,\ldots)$, $a_i \leq a_j'$ $(i \neq j,\ i,j = 1,2,\ldots)$ then $\bigvee_{i=1}^{\infty} a_i$ exists in L.

Note that $x \vee y$, $x \wedge y$ denote the l.u.b (g.l.b) of x,y, if they exist. Further we write $x \perp y$ if $x \leq y'$ saying that x,y are orthogonal. Thus (v) means that L is closed under countable orthogonal unions, while (iv) asserts that $c \perp a$. It may be proved that $c = b \wedge a'$.

Two elements $a, b \in L$ are said to be compatible (notation $a \leftrightarrow b$) if there exist mutually orthogonal elements u,v,z such that $a = u \vee z$, $b = v \vee z$. A set $A \subset L$ is said to be compatible if $a \leftrightarrow b$ for any $a, b \in L$. We say that A is strongly compatible if for any $a, b \in L$ there exist u,v,z in the smallest logic L_0 generated by A such that $a = u \vee z$, $b = v \vee z$. In case u,v,z may be chosen such that they belong to A, we say that A is inner compatible.

If L is a logic and $z: \mathcal{B} \longrightarrow L$ is a mapping defined on the collection $\mathcal{B}$ of all Borel sets of the real line R such that

(i) $z(R) = 1$,

(ii) if $E, F \in \mathcal{B}$, $E \cap F = \emptyset$ then $z(E) \perp z(F)$,

(iii) $z(\bigcup_{n=1}^{\infty} E_n) = \bigvee_{n=1}^{\infty} z(E_n)$ for any sequence of sets E_n such that $E_i \cap E_j = \emptyset$ $(i \neq j,\ i,j = 1,2,\ldots)$, then z is said to be an observable on L.

The compatibility and strong compatibility of a collection $\{z_t : t \in T\}$ of observables is defined as compatibility and strong compatibility of the collection $\{z_t(B) : t \in T,\ B \in \mathcal{B}\}$ respectively.

The results of this paper will merely concern special types of logics, sometimes called σ-classes (see [2]). Since these special

---

[1)]Department of Probability and Statistics, Komensky University, Bratislava, Czechoslovakia.

classes are generalizations of Boolean σ-algebras and since the logics themselves are related to quantum mechanics, we shall use the name q-algebra instead of σ-class.

A q-algebra is a nonempty collection $\mathcal{Q}$ of subsets of a set S ($\mathcal{Q} \subset \mathcal{P}(S)$, where $\mathcal{P}(S)$ is the potence set of S) such that

(i) $E_n \in \mathcal{Q}$, $n = 1,2,\ldots$, $E_i \cap E_j = \emptyset$ $(i \neq j,\ i,j = 1,2,\ldots)$ implies $\bigcup_{i=1}^{\infty} E_i \in \mathcal{Q}$,

(ii) $E,F \in \mathcal{Q}$, $E \supset F$ implies $E - F \in \mathcal{Q}$.

In fact a q-algebra $\mathcal{Q}$ is a logic, where the partial order is defined by means of the inclusion and the complementation is the set-theoretical complementation.

Note that in a q-algebra $\mathcal{Q}$ two elements E,F are compatible exactly if $E \cap F \in \mathcal{Q}$.

We shall use the notation $q(\mathcal{Z})$ for the smallest q-algebra containing a collection $\mathcal{Z}$.

On a q-algebra $\mathcal{Q} \subset \mathcal{P}(S)$ we can obtain a special kind of observable by means of a real function $f: S \to R$. In fact any such function induces a mapping $f^{-1}: \mathcal{B} \to \mathcal{P}(S)$ defined by means of the inverse image $f^{-1}(E)$ where $E \subset R$. If $f^{-1}(E) \in \mathcal{Q}$ for any $E \in \mathcal{B}$, then f^{-1} is an observable. It is called an observable induced by f.

Given two q-algebras $\mathcal{Q}_1$, $\mathcal{Q}_2$ ($\mathcal{Q}_1 \subset \mathcal{P}(S)$, $\mathcal{Q}_2 \subset \mathcal{P}(U)$), we define their product $\mathcal{Q}_1 \otimes \mathcal{Q}_2$ to be the smallest q-algebra of $S \times U$ generated by the collection of all $E \times F$ where $E \in \mathcal{Q}_1$, $F \in \mathcal{Q}_2$.

The usual measure-theoretical product of two σ-algebras $\mathcal{Q}_1$, $\mathcal{Q}_2$ will be denoted $\mathcal{Q}_1 \times_\sigma \mathcal{Q}_2$.

In general if we have two q-algebras $\mathcal{Q}_1 \subset \mathcal{P}(S)$, $\mathcal{Q}_2 \subset \mathcal{P}(U)$ and a function $f: S \times U \to R$, the sections f_s, f^u ($f_s(u) = f(s,u)$, $u \in U$; $f^u(s) = f(s,u)$, $s \in S$) may induce observables on $\mathcal{Q}_2$ and $\mathcal{Q}_1$ respectively for any $s \in S$ and $u \in U$ and f need not induce an observable on $\mathcal{Q}_1 \otimes \mathcal{Q}_2$.

In fact, it is sufficient to consider the example of two measurable spaces $(X, \mathcal{S})$, $(Y, \mathcal{T})$ and a function $f: X \times Y \to R$ such that f_x, f^y are measurable and f is not $\mathcal{S} \times_\sigma \mathcal{T}$-measurable. Such a function can be easily constructed (see [4], p. 142, exercise (2)). Since $\mathcal{S}$, $\mathcal{T}$ are evidently q-algebras and $\mathcal{S} \otimes \mathcal{T} \subset \mathcal{S} \times_\sigma \mathcal{T}$, we have that f_x, f^y induce observables on $\mathcal{T}$, $\mathcal{S}$ respectively, but f does not induce an observable on $\mathcal{S} \otimes \mathcal{T}$.

There are well known results in classical measurable spaces asserting that under certain additional conditions measurability of sections implies the measurability of f. We prove that some of such results may be proved for observables, if we consider a product $\mathcal{S} \otimes \mathcal{Q}$, where $\mathcal{S}$ is a σ-algebra of sets in a suitable topological space and $\mathcal{Q}$ is a q-algebra of an abstract set.

The following lemma follows for q-algebras from a result proved in [5] (see [3], [6] for general logics).

Lemma 1. Let L be a logic. Let $A \subset L$ be strongly compatible. Then the logic L_0 generated by A is a Boolean σ-algebra.

Lemma 2. Let L be a logic. Let $A \subset L$ be inner compatible. Then the logic L_0 generated by A is a Boolean σ-algebra.

Proof. It follows from Lemma 1.

Lemma 3. Let $\{z_t\}$, $(t \in T)$, be a collection of observables on a q-algebra $\mathcal{Q}$. Let $\mathcal{Z} = \{z_t(B):\ t \in T,\ B \in \mathcal{B}\}$. Let $\mathcal{C} \subset \mathcal{B}$ be an inner compatible collection generating $\mathcal{B}$ and such that for any C,D belonging to $\mathcal{C}$ and any t_1, $t_2 \in T$ always $z_{t_1}(C) \cap z_{t_2}(D) \in q(\mathcal{Z})$. Then $\mathcal{Z}$ is strongly compatible.

Proof. We have to prove

$$z_{t_1}(A) \cap z_{t_2}(B) \in q(\mathcal{Z}) \tag{1}$$

for any t_1, $t_2 \in T$ and any $A, B \in \mathcal{B}$. Let $C \in \mathcal{C}$, t_1, $t_2 \in T$ be fixed. Put

$$\mathcal{S}(C) = \{E:\ E \in \mathcal{B},\ z_{t_1}(C) \cap z_{t_2}(E) \in q(\mathcal{Z})\}.$$

One can easily verify that $\mathcal{S}(C)$ is a q-algebra. Since $\mathcal{S}(C) \supset \mathcal{C}$ we have $\mathcal{S}(C) \supset q(\mathcal{C})$. But $q(\mathcal{C}) = \mathcal{B}$ (see Lemma 2). Thus $\mathcal{S}(C) \supset \mathcal{B}$. Since $\mathcal{S}(C) \subset \mathcal{B}$ is evident, we have $\mathcal{S}(C) = \mathcal{B}$.

Now let $B \in \mathcal{B}$ be fixed. Put

$$\mathcal{S}(B) = \{E:\ E \in \mathcal{B},\ z_{t_1}(E) \cap z_{t_2}(B) \in q(\mathcal{Z})\}. \tag{2}$$

According what was proved, we have $\mathcal{S}(B) \supset \mathcal{C}$. Again it is easy to prove that $\mathcal{S}(B) = \mathcal{B}$. This and (2) imply (1).

Lemma 4. Let T be a first countable topological space and $\mathcal{Q}$ a q-algebra of subsets of X. Let $f: T \times X \to R$. Denote by z_t $(t \in T)$ the mappings defined on $\mathcal{B}$ such that

$$z_t(B) = \{x:\ f(t,x) \in B\}$$

for every $B \in \mathcal{B}$. Let f be continuous on T for every $x \in X$. Let there exist a dense set $D \subset T$ such that $z_t(B) \in \mathcal{Q}$ for every $t \in D$, $B \in \mathcal{B}$ and let $\{z_t(B):\ B \in \mathcal{B},\ t \in D\}$ be strongly compatible. Then $\{z_t:\ t \in T\}$ is a strongly compatible collection of observables.

Proof. Due to Lemma 3 it is sufficient to prove that for an inner compatible collection $\mathcal{C}$ generating $\mathcal{B}$ and for $C, D \in \mathcal{C}$ and any $t, r \in T$ we have

$$z_t(C) \cap z_r(D) \in q(\mathcal{Z}) \tag{3}$$

where $\mathcal{Z} = \{z_t(B): t \in T, B \in \mathcal{B}\}$.

Take $\mathcal{C}$ as the collection of all open intervals. Choose sequences $\{t_n\}, \{r_n\}$ $(n = 1,2,\ldots)$ respectively such that

$$t = \lim t_n, \quad r = \lim r_n, \quad t_n, r_n \in D.$$

Further let $(a_k, b_k), (c_k, d_k)$ $(k = 1,2,\ldots)$ be intervals such that

$$(a, b) \supset (a_k, b_k), \; (c, d) \supset (c_k, d_k),$$

$\lim a_k = a, \; \lim b_k = b, \; \lim c_k = c, \; \lim d_k = d.$

Using the continuity of $f(t,x)$ for every $x \in X$, we obtain

$$z_t((a,b)) = \bigcup_{k=1}^{\infty} \bigcap_{n=1}^{\infty} \bigcup_{m=n}^{\infty} z_{t_m}((a_k,b_k)) \tag{4}$$

$$z_r((c,d)) = \bigcup_{k=1}^{\infty} \bigcap_{n=1}^{\infty} \bigcup_{m=n}^{\infty} z_{r_m}((c_k,d_k)) \tag{5}$$

Since the collection $\mathcal{Z}_1 = \{z_t(B): t \in D, B \in \mathcal{B}\}$ is strongly compatible we obtain (Lemma 1) that $q(\mathcal{Z}_1)$ is a σ-algebra. Since $z_{t_m}((a_k,b_k)), z_{r_m}((c_k,d_k))$ belong to the σ-algebra $q(\mathcal{Z}_1) \subset \mathcal{Q}$, from (3), (4) we have $z_t((a,b)) \in q(\mathcal{Z}_1)$, $z_r((a,b)) \in q(\mathcal{Z}_1)$. Thus $z_t((a,b)) \cap z_r((a,b)) \in q(\mathcal{Z}_1) \subset q(\mathcal{Z})$. Thus (3) holds and the proof is finished.

Using Lemma 4 we can apply some methods for proving measurability of functions of two variables (see [1]) to obtain the following

Theorem 1. Let T be a second countable topological space and $\mathcal{S}$ a σ-algebra of subsets of T containing all open sets. Let $\mathcal{Q}$ be a q-algebra of subsets of X. Let $f: T \times X \to R$ be such that f^x is continuous for every x and there exists a dense set $D \subset T$ such that the mappings z_t induced by f_t for any $t \in D$ are strongly compatible observables on $\mathcal{Q}$. Then f induces an observable on the product $\mathcal{S} \otimes \mathcal{Q}$

Proof. According to Lemma 4 the function f induces for any $t \in T$ an observable z_t on $\mathcal{Q}$ and the collection $\{z_t: t \in T\}$ is strongly compatible. Thus there exists (see Lemma 1) a σ-algebra $\mathcal{S}_1 \subset \mathcal{Q}$ containing the collection $\{z_t(B): B \in \mathcal{B}, t \in T\}$. Using a result from [1] ([1], Lemma 4.2) there exists a collection $\{P_n^k\}$, $k = 1,2,\ldots;$ $n \in N_k$, where N_k is either the set of all positive integers or $N_k = \{1,2,\ldots,i_k\}$, such that $P_n^k \in \mathcal{S}$, $P_n^k \cap P_m^k = \emptyset$ if $m \neq n$ and

$\bigcup_n P_n^k = T$ for $k = 1,2,\dots$. Moreover the sets P_n^k may be chosen such that for any open set G and any $t \in G$ there exists k_o such that if $k > k_o$ then $t \in P_n^k$ implies $P_n^k \subset G$.

Choose a point t_n^k in each P_n^k. Define a sequence $\{f_k\}_{k=1}^{\infty}$ of functions f_k: $T \times X \to R$ by

$$f_k(t,x) = f(t_n^k, x) \text{ if } t \in P_n^k.$$

For any $B \in \mathcal{B}$ we have

$$f_k^{-1}(B) = \bigcup_{n=1}^{N_k} P_n^k \times z_{t_n^k}(B) \tag{6}$$

It follows from (6) that $f_k^{-1}(B)$ belongs to the σ-algebra generated by sets $E \times z_t(B)$ where $E \in \mathcal{S}$, $t \in T$, $B \in \mathcal{B}$. Hence $f_k^{-1}(B)$ belongs to the product-σ-algebra $\mathcal{S} \times_\sigma \mathcal{S}_1$. We prove that

$$\mathcal{S} \times_\sigma \mathcal{S}_1 = \mathcal{S} \otimes \mathcal{S}_1 . \tag{7}$$

To prove the last, it suffices to show (see Lemma 1) that the collection of all $E \times F$ where $E \in \mathcal{S}$, $F \in \mathcal{S}_1$ is strongly compatible. Taking $E_1 \times F_1$, $E_2 \times F_2$, where $E_i \in \mathcal{S}$, $F_i \in \mathcal{S}_1$ $(i = 1,2)$, we have $(E_1 \times F_1) \cap (E_2 \times F_2) = (E_1 \cap E_2) \times (F_1 \cap F_2)$. Since $E_1 \cap E_2 \in \mathcal{S}$, $F_1 \cap F_2 \in \mathcal{S}_1$, the inner compatibility and hence the strong compatibility is proved. Thus (7) is proved and we have

$$\mathcal{S} \times_\sigma \mathcal{S}_1 = \mathcal{S} \otimes \mathcal{S}_1 \subset \mathcal{S} \otimes Q \tag{8}$$

The functions f_k being $\mathcal{S} \times_\sigma \mathcal{S}_1$-measurable, we have that $\lim f_k$, if it exists, is $\mathcal{S} \times_\sigma \mathcal{S}_1$-measurable. Using the relation of P_n^k to open sets and the continuity of f in the variable t, one can prove that $\lim f_k = f$. Hence f is $\mathcal{S} \times_\sigma \mathcal{S}_1$-measurable. This and the inclusion (8) imply that f induces in $\mathcal{S} \otimes Q$ an observable.

A more general version of Theorem 1 may be proved using the same method as in Theorem 1. To formulate such theorem we mention some notions introduced in [1] .

If $(T, \mathcal{S})$ is a measurable (not necessary a topological) space, then a collection $\{P_n^k\}$ of nonempty sets such that $P_n^k \cap P_m^k = \emptyset$ if $m \neq n$ $(m, n = 1,2,\dots , k = 1,2,\dots , P_n^k \in \mathcal{S} , \bigcup_{n \in N_k} P_n^k = T$ for $k = 1,2,\dots)$ is said to be a $\mathcal{P}$-system on X.

Given a set T and a $\mathcal{P}$-system $\mathcal{P}$ on T a function g: $T \to R$ is said to be regular at t_o with respect to $\mathcal{P}$ if for every open $V \subset R$ with $g(t_o) \in V$ there exists k_o such that if $k > k_o$, then $t, t_o \in P_n^k$ imply $f(t) \in V$. The function g is said to be relatar with respect to $\mathcal{P}$, if it is regular with respect to $\mathcal{P}$ at any $t \in T$.

Now we are able to formulate the following

Theorem 2. Let $(T, \mathcal{S})$ be a measurable space. Let $\mathcal{Q}$ be a q-algebra of subsets of a set X. Let a $\mathcal{P}$ on T exist and let f: $T \times \mathcal{S} \longrightarrow R$ be such that for any $x \in X$ $f(t,x)$ is regular with respect to $\mathcal{P}$ and the mappings z_t induced by the sections f_t form a strongly compatible collection of observables on $\mathcal{Q}$. Then f induces an observable on $\mathcal{S} \otimes \mathcal{Q}$.

Concluding remarks. One can easily obtain Theorem 1 from Theorem 2. It is sufficient to know, some facts about existence of $\mathcal{P}$-systems on second countable topological spaces (see [1]). Further some corollaries may be obtained from Theorem 2 for functions f: $T \times X \rightarrow R$, the sections f^x of which satisfy suitable conditions for constructing a $\mathcal{P}$-system with respect to which f^x are regular. The assumptions on f_t remain unchanged. Thus the situation for the "product observability" under the strong compatibility of f_t $(t \in T)$ resembles the situation of the product measurability.

References

[1] Dravecký, J. - Neubrunn, T.: Measurability of functions of two variables. Mat. Čas. 23, (1973), 147-157.

[2] Gudder, S.P.: Quantum probability spaces. Proc. Amer. Math. Soc. 21, (1969), 296-302.

[3] Guz, W.: Quantum logic and a theorem on commensurability. Reports on mathematical physics, 2, (1971), 53-61.

[4] Halmos, P.R.: Measure theory. New York, 1950.

[5] Neubrunn, T.: A note on quantum probability spaces. Proc. Amer. Math. Soc. 25, (1970), 672-675.

[6] Neubrunn, T.: On certain generalized random variables. Acta Fac. RNUC, 29, (1974), 1-6.

[7] Varadarajan. V.S.: Probability in physics and a theorem on simultaneous observability. Comm. in pure and applied Math. 15, 1962, 189-217, correction, loc. cit. 18, (1965).

TWO CLASSIFICATION PROBLEMS FOR FINITE STATE MARKOV CHAINS

William Parry[1)] and Selim Tuncel[1)]

0. Introduction

The classification problem for finite state Markov chains has many versions. The most abstract version which treats the chains merely as invertible measure-preserving transformations was solved by Friedman and Ornstein [2]. The finitary version, which takes into account the state structure of the chains was solved by Keane and Smorodinsky [3]. We wish to speak about two other versions. In the first place we shall concentrate on block-isomorphisms. These are essentially measure-preserving homeomorphisms which conjugate shifts. Consequently, we have to take into account both the measure-structure and the topological structure. As far as the latter is concerned the problem amounts to that investigated by Williams in his paper on the classification of shifts of finite type [10]. Williams showed that a topological conjugacy between two irreducible shifts of finite type exists if and only if their defining matrices are strong shift equivalent, a condition which implies shift equivalence which, in turn, implies identical zeta functions. A mistake in his paper left open (even to the present time) the question as to whether shift equivalence implies topological conjugacy. In a successful attempt to simplify the topological classification of shifts of finite type Adler and Marcus [1] showed that topological entropy and period, classified such shifts from the point of view of almost topological conjugacy. This notion brings us to the second of our themes as there is an analogous concept applicable to (stochastic) Markov chains. The results to be discussed will appear in [5] , [6].

1. Block-isomorphism

Let P be a $k \times k$ irreducible stochastic matrix and let P^t be the matrix obtained from P by raising each entry to the power of t, so that, for example, P^o is a zero-one matrix. Let $X_P = X_{P^o}$ be the space of doubly infinite sequences $x = \{x_n\}$ chosen from $1,2,\dots,k$ such that $P^o(x_n,x_{n+1}) = 1$. X_P is a zero-dimensional compact space with the Tychorov topology. The shift $T_P = T_{P^o}$, which is defined by $T_P\{x_n\} = \{x_{n+1}\}$, preserves the Markov measure m_P defined by P in the usual way. We refer to (T_P,m_P) or even T_P as a *Markov shift* (or Markov chain.)

If P,Q are two irreducible stochastic matrices we say that T_P, T_Q are *block-isomorphic* if there exists a measure-preserving homeomorphism ϕ of X_P onto X_Q such that $\phi T_P = T_Q\phi$.

In [7] Parry and Williams proved that T_P, T_Q are block-isomorphic if and only if there exist parallel series of matrix equations

$$(1.1)\qquad \begin{array}{l} P = U_1V_1,\ V_1U_1 = U_2V_2,\ \dots\ V_nU_n = Q \\ P^o = U_1^oV_1^o,\ V_1^oU_1^o = U_2^oV_2^o,\ \dots\ V_n^oU_n^o = Q^o \end{array}$$

where U_i, V_i are stochastic rectangular. One can even insist that the second line should consist entirely of zero-one matrices. This being the case it is not hard to see that

1) Mathematics Institute, University of Warwick, Coventry CV4 7AL

the parallel series can be written as one series

(1.2) $P^t = U_1(t)V_1(t),\ V_1(t)U_1(t) = U_2(t)V_2(t),\ \ldots\ V_n(t)U_n(t) = Q^t$

for all real t, where $U_i(t)$, $V_i(t)$ are matrices over the semi-ring $\mathbb{Z}^+(\exp) \subset \mathbb{Z}(\exp)$. ($\mathbb{Z}^+(\exp)$ denotes non-negative integral combinations of exponential functions e^{at}, a real.)

The condition (1.2) we call *strong shift-equivalence* (for the stochastic setting), and is pictured in the following diagram:

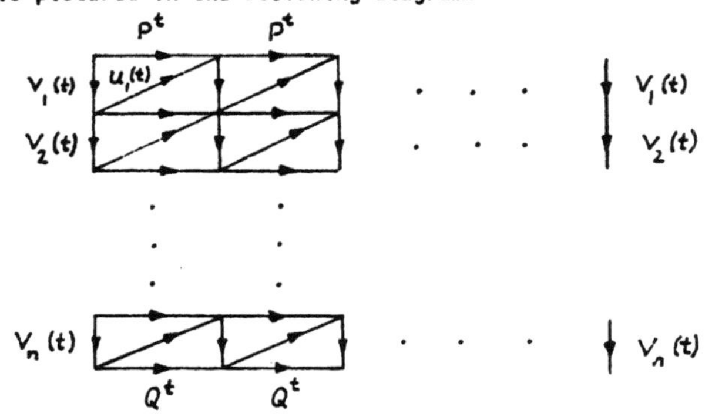

Defining $V(t) = V_n(t) \ldots V_2(t)V_1(t)$, $U(t) = U_1(t) \ldots U_n(t)$ we have

(1.3) $V(t)P^t = Q^tV(t),\ P^tU(t) = U(t)Q^t,\ U(t)V(t) = (P^t)^{(n)}$
$\qquad V(t)U(t) = (Q^t)^{(n)}$

which we call *shift-equivalence* (for the stochastic setting.)

One should note that if we put t = 0 we have Williams's definitions of strong shift equivalence and shift equivalence for the purely topological problem.

The following definition of a zeta function suggests itself:

$$(1.4)\quad \zeta(s,t) = \frac{1}{\det(I-sP^t)} = \exp \sum_{n=1}^{\infty} \frac{s^n}{n} \operatorname{Tr}(P^t)^{(n)} = \exp \sum_{n=1}^{\infty} \frac{s^n}{n} \sum_{T_p^n x = x} \exp t\{I_{T_p}(x) + \ldots + I_{T_p}(T_p^{n-1}x)\}$$

It is not hard to see that this is an invariant of shift-equivalence and therefore of block-isomorphism. The maximum root of $\det(sI-P^t)$ (i.e. the maximum eigenvalue of P^t) which we denote by $\beta(t)$ is, therefore, also an invariant.

Let P_t denote the stochasticisation of P^t i.e. $P_t = (r(t)^{-1}P^t r(t)/\beta(t))$ where r(t) is the diagonal matrix obtained from the right eigenvector corresponding to $\beta(t)$. Let m_{P_t} denote the corresponding Markov probability, and note that the one-parameter multiplicative group consisting of the positive reals $\mathbb{R}^+$ acts on all Markov probabilities m supported on X_{P^0} by $t{:}m_P \to m_{P_t}$. Except for the maximum entropy measure, each orbit is a copy of $\mathbb{R}^+$. (No fixed points, no closed orbits,) The invariant measure m_{P_0} with maximum entropy on X_{P^0} is fixed by this action.

If ϕ is a block-isomorphism from (X_P, T_P) to (X_Q, T_Q) then

(1.5) $I_P = I_Q \circ \phi + f \circ T_P - f$

for some continuous f where $I_P = I(\alpha | T_P^{-1}\alpha)$ is the information function of T_P. Multiplying this equation by t gives

$$I_{P_t} = I_{Q_t} \circ \phi + f_t \circ T_{P_t} - f_t$$

for a continuous function f_t, and this equation ensures that ϕ is also a block isomorphism from (X_{P_t}, T_{P_t}) to (X_{Q_t}, T_{Q_t}), for all $t \in (\mathbb{R}^+, 0)$. Thus ϕ induces a map from the orbit $\{m_{P_t}\}$ to the orbit $\{m_{Q_t}\}$. This shows why the parameterised notion of strong shift equivalence is natural.

2. The function $\beta(t)$.

Before proceeding to our second problem we make some observations concerning the function $\beta(t)$.

Topological entropy is easily computed to be $\log \beta(0)$. As for measure-theoretic entropy $h(m_P)$, this together with $h(m_{P_t})$ can be computed from the equation

$$\beta(m, t_s) = \beta(m,t)^s \, \beta(m_t, s)$$

where $m_t = m_{P_t}$, $m_1 = m$, and $\beta(m_t, s)$ is the maximum eigenvalue of $(P_t)^s$. One also uses the matrix eigenvector equation and obtains

$$h(m_t) = \log \beta(m,t) - t(\beta'(m,t)/\beta(m,t)),$$

so that $h(m_0) = \log \beta(m,0)$, $h(m) = -\beta'(m,1)$. Of course, $\beta(m,1) = 1$.

There is another invariant which is worth computing, which appears whenever we have the equation (1.5). This is defined as

$$\sigma^2(T_P) = \lim_{n\to\infty} \frac{1}{m} \int |I_P + \dots + I_P \circ T_P^{n-1} - nh(m)|^2 dm.$$

Again, this can be computed from $\beta(t)$:

$$\sigma^2(T_{P_t}) = t^2(\beta''(m,t)/\beta(m,t)) - (\log \beta(m,t) - h(m_t))^2$$

so that

$$\sigma^2(T_{P_0}) = 0$$

and

$$\sigma^2(T_P) = \beta''(m,1) - (h(m))^2.$$

3. Finite equivalence

In the spirit of Adler and Marcus we may define two Markov shifts T_P, T_Q to be *almost block-isomorphic* if there exists another Markov shift T_R and two continuous measure-preserving surjections ϕ,ψ such that $\phi T_R = T_P\phi$, $\psi T_R = T_Q\psi$ and ϕ,ψ are a.e. one-one.

A weaker requirement is *finite equivalence*. This amounts to the above definition with "a.e one-one" replaced by "bounded to one".

If $\phi T_R = T_P\phi$ where ϕ is a continuous bounded to one measure-preserving surjection then we have the equation

$$I_R = I_P o \phi + f \circ T_R - f \tag{3.1}$$

where f is continuous. In case ϕ is a.e. one-one the group

$$\Lambda_R = \{(a,b): \exp i(a + bI_R) = F \circ T_R/F \text{ for some measurable } F \text{ such that } |F| = 1\}$$

coincides with Λ_P, as may be checked using (3.1). Hence Λ_P is an invariant of almost block-isomorphism.

One can evaluate $\beta_P(t) \equiv \beta(m_P,t) = \beta(t)$ as follows

$$\lim_{n\to\infty} \frac{1}{n} \log \int \exp t(I_P + \dots + I_P \circ T_P^{n-1})dm_P = \underline{P}((t-1)I_P) = \beta_P(t-1)$$

where $\underline{P}$ denotes pressure. (cf. [8] [9] From this fact and (3.1) it is immediate that $\beta_P(t) \equiv \beta_Q(t)$ when T_P, T_Q are finitely equivalent.

It is natural to wonder whether identical β functions and identical periods imply that two Markov shifts are almost block-isomorphic. This is *not* the case as can be seen from the examples

$P = \begin{pmatrix} p & q \\ p & q \end{pmatrix}$, $Q = \begin{pmatrix} p & q \\ q & p \end{pmatrix}$ $(p,q \neq \frac{1}{2})$. Certainly $\beta_P(t) = \beta_Q(t) = p^t + q^t$, whereas

$\Lambda_P \neq \Lambda_Q$.

Thus the exact analogue of a theory a la Adler and Marcus is false. Instead we are concerned with the *Conjecture* If $\beta_P(t) = \beta_Q(t)$ then T_P, T_Q are finitely equivalent. Such a result would generalise the main theorem of [4] which is concerned with Markov measures of maximal type.

As in 1, let $\mathbb{Z}(\exp)$ denote the ring of integral combinations of exponential functions e^{at}, a real, and let $\mathbb{Z}^+(\exp)$ denote the sub-semi-ring of such combinations with *non-negative* coefficients. Generalising a result of Furstenberg we can prove

Theorem. If P,Q are irreducible stochastic matrices with identical β functions then there exists a non-trivial matrix U(t) with entries in $\mathbb{Z}(\exp)$ such that

$$U(t)P^t = Q^tU(t)$$

The converse is also true.

(By non-trivial, we mean no rows or columns are trivial.) *Moreover for each* $t_0 \in \mathbf{R}$, $U(t)$ *can be chosen so that* $U(t_0)$ *is a strictly positive matrix.*

Problem 1. Can $U(t)$ be chosen so that its entries belong to $\mathbf{Z}^*(\exp)$, the sub-semi-ring of $\mathbf{Z}(\exp)$ consisting of functions nowhere negative?

Problem 2. Can $U(t)$ be chosen so that its entries belong to $\mathbf{Z}^+(\exp)$?

Our success with many examples leads us to conjecture, tentatively, the affirmative to problem 2. In the absence of a proof, we replace the hypothesis $\beta_P(t) = \beta_Q(t)$ by the hypothesis

(3.2) $$U(t)P^t = Q^t U(t)$$

where $U(t)$ is a non-trivial matrix over $\mathbf{Z}^+(\exp)$.

Theorem. Given (3.2) there is a commutative diagram

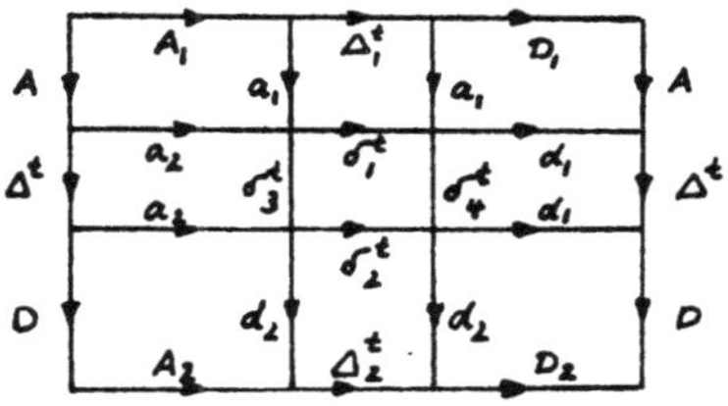

where $P^t = D_1\Delta_1^t A_1$, $Q^t = D_2\Delta_2^t A_2$, $U(t) = D\Delta^t A$ and where all A's, a's are *amalgamations* (each row is a unit vector and no columns are trivial) all D's, d's are *divisions* (transpositions of amalgamations) and all Δ's, δ's are positive diagonal matrices. (As usual, the exponent t indicates entries have been raised to this power). The proof of this theorem depends on the following:

Lemma. If $U(t)$ is a non-trivial rectangular matrix over $\mathbf{Z}^+(\exp)$ then there is a decomposition

$$U(t) = D\Delta^t A.$$

Moreover this decomposition is essentially unique in the sense that if $U(t) = D_1\Delta_1^t A_1$ then there is a permutation σ such that $D_1 = D\sigma$, $A_1 = \sigma^{-1}A$ and $\Delta_1 = \sigma^{-1}\Delta\sigma$.

At this point perhaps it is worth commenting that the lemma enables us to interpret any square matrix $P(t)$ over $\mathbf{Z}^+(\exp)$ such that $P(1)$ is stochastic, as the defining matrix of a Markov chain. For we may write $P(t) = D\Delta^t A$ and define $Q = AD\Delta, Q^t = AD\Delta^t$ so that $P(t), Q^t$ are one-step strong-shift equivalent. Thus the 1×1 matrix $(p_1^t + \ldots + p_k^t)$ represents the Bernoulli shift with weights $p_1, \ldots, p_k$.

If we assume, as we may, that $U(t)$ in (3.2) is minimal in the sense that if $V(t)$ is another candidate with $U(t) - V(t)$ positive (i.e. has entries in $\mathbf{Z}^+(\exp)$) then $U(t) = V(t)$, then the matrices $d_1\,\delta_1^t\,a_2$, $d_1\delta_2^t\,a_2$ are irreducible. This is not immediate but it is not difficult to prove. We can write $R^t = d_1\delta_1^t a_2$ and $S^t = d_1\delta_2^t a_2$ for certain positive matrices R, S and in fact R is stochastic.

From the above diagram we conclude the following commutative diagrams:

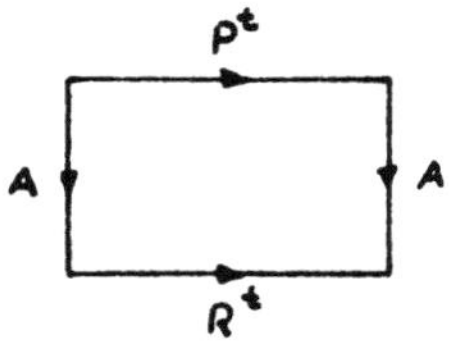

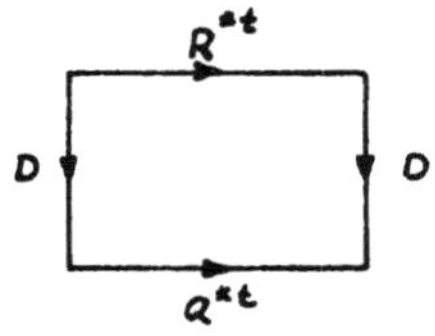

where * denotes transposition.

The first of these implies immediately that $\phi T_R = T_P\phi$ for the bounded to one continuous measure-preserving map ϕ defined by $\phi(x) = y$ when $A(x_n, y_n) = 1$. In the same way the second of these implies that $\psi T_R^{-1} = T_Q^{-1}\psi$ for the bounded to one continuous measure-preserving map ψ defined by $\psi(x) = z$ when $D(z_n, x_n) = 1$. In this way we establish

<u>Theorem</u> Given (3.2), the Markov shifts T_P, T_Q are finitely equivalent.

To summarise: If $\beta_P(t) \equiv \beta_Q(t)$ then we can establish the existence of a matrix $U(t)$ over $Z(\exp)$ such that $U(t)P^t = Q^tU(t)$. However, as yet we are unable to show that the entries can be taken as elements of $Z^+(\exp)$. If, however, this is assumed then T_P, T_Q are finitely equivalent.

As an illustration let us show the following:

<u>Theorem.</u> Let P be an irreducible $\ell \times \ell$ stochastic matrix all of whose rows are permutations of the numbers $p_1, p_2, \ldots, p_k, 0, \ldots 0$, ($k \leq \ell$ fixed,) where $p_1, \ldots, p_k$ are strictly positive. Let Q be the Bernoulli stochastic matrix $\begin{pmatrix} p_1 & \cdots & p_k \\ \cdots & \cdots & \cdots \\ p_1 & \cdots & p_k \end{pmatrix}$.

Then T_P, T_Q are finitely equivalent.

<u>Proof</u>

Evidently $P^t.1 = (p_1^t + \ldots + p_k^t).1$ where 1 denotes the column vector (or matrix) consisting of 1's. Now write $P^t = D\Delta^tA$ and note that 1 is an amalgamation, and

$$(p_1^t + \ldots + p_k^t) = (1 \ldots 1)\begin{pmatrix} p_1^t & & \\ & \ddots & \\ & & p_k^t \end{pmatrix}\begin{pmatrix} 1 \\ \vdots \\ 1 \end{pmatrix}$$

Evidently we have

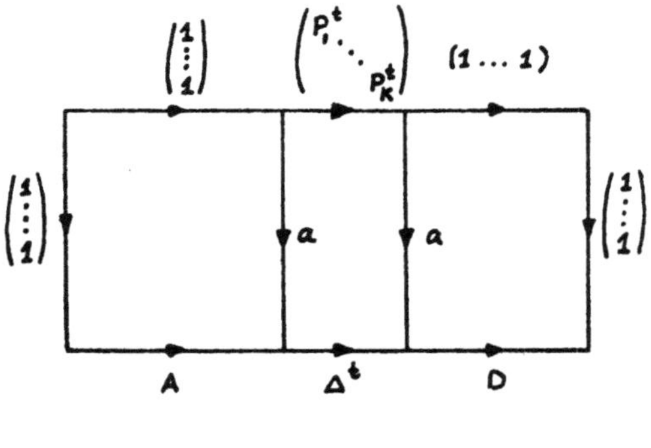

Rearranging this diagram we have

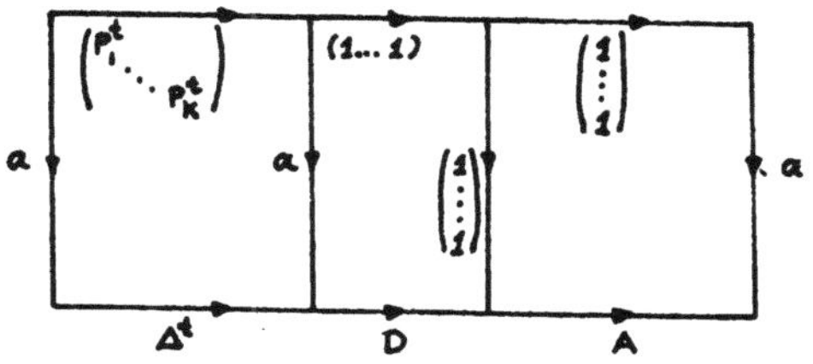

i.e. a $Q^t = (AD\Delta^t)a$.

But the shift based on $AD\Delta$ is one-step equivalent to the shift based on $P = D\Delta A$. Thus T_P, T_Q are finitely equivalent. In fact, T_Q is a bounded to one factor of T_P.

References

1. R.L. Adler & B. Marcus., Topological entropy and equivalence of dynamical systems. Mem. A.M.S. 219 (1979).
2. N.A. Friedman & D.S. Ornstein., On isomorphisms of weak Bernoulli transformations. Adv. in Math. 5 (1970), 365-394.
3. M. Keane & M. Smorodinsky., The finitary isomorphism theorem for Markov shifts. (To appear)
4. W. Parry., A finitary classification of topological Markov chains and sofic systems, Bull. A.M.S. 9 (1977) 86-92.
5. W. Parry & S. Tuncel., On the classification of finite Markov chains by finite equivalence. To appear in Ergodic theory and Dynamical Systems.
6. W. Parry & S. Tuncel., On the topological and stochastic structure of Markov chains. To appear in Bull. L.M.S.
7. W. Parry & R.F. Williams., Block codes and a zeta function for finite Markov chains. Proc. L.M.S. 35 (1977) 483-495.
8. D. Ruelle., Thermodynamic formalism. Addison-Wesley, Reading Mass., 1978.
9. S. Tuncel., Conditional pressure and coding, Israel J. Math. 39 (1981), 101-112.
10. R.F. Williams., Classification of subshifts of finite type, Ann. Math. 98 (1973) 120-153; Errata Ann. Math. 99 (1974) 380.

ON STOCHASTIC PROPERTIES OF ATTRACTORS

Ya. B. Pesin[1)]

1. Introduction.

1. One of the most important achievements in the theory of smooth dynamical systems is the discovery of such motions in the deterministic dynamical systems, which have a purely stochastic behaviour, i.e. which can be considered as typical realizations of a random process. This fact was already mentioned in the thirties by Hopf and Hedlund in their studies of statistics of geodesics on negative curvature surfaces. However the exact formulations and results appeared only in the sixties in the papers of Anosov and Sinaĭ [1], [2], [14]. These papers started the new direction, which is often called now as hyperbolic theory (cf. [11]). It was found that quite an universal way of the appearance of stochasticity in the deterministic system is some unstability of its orbits which can be expressed with the help of the so-called hyperbolicity conditions. They are formulated in the terms of the variation equations, i.e. by means of some conditions on the differentials of dynamical system along on individual orbit (in detail, cf. below). At present four types of the hyperbolicity conditions are wellknown. They are as follows: 1) uniform and total; 2) uniform and partial; 3) non-uniform and total; 4) non-uniform and partial hyperbolicity. The first case was considered in the original papers of Anosov and Sinaĭ. Corresponding dynamical systems were called Anosov systems. On the one hand the rigidity of the requirements of uniform and total hyperbolicity opened a possibility for a thorough study of statistic properties of Anosov systems and on the other hand it bounded the possible applications. It is known that Anosov systems are not connected directly with any physical systems, and the verication of the conditions of uniform and total hyperbolicity is, as a rule, very difficult. May be for this reason at first the hyperbolic theory did not get due acknowledgement among physicists.

2. However the situation changed radically in the seventies. First of all other weaker hyperbolicity conditions were investigated. It turned out that , for example , one can easily examine the non-uniform hyperbolicity conditions counting up the Ljapunov characteristic exponents along "typical" orbits. The corresponding methods of the calcu-

1) All-Union Research Institute on Metrological Survey ,
1 , Ezdakov Pereulok , Moscow 117334 , USSR

lation using the computer are wellknown at present (at least in two-dimensional case). These methods open a possibility of stating on instability of the orbits (which is, as a rule, rather weak) in some physical models.

We should say about the great success achieved by Bunimovich and Sinaĭ (cf. [6], [5], [8]) in the investigation of the statistic properties of billiards dynamical systems, in which the weakening of the hyperbolicity takes place at the expense of the special set where the map or its differential are discontinuous.

3. Thus new objects arose in the mathematical investigation of some physical models and at the first place the model of the appearance of turbulence, suggested by Ruelle and Takens (cf. [13]) belong to them. They are invariant attracting sets of a special type which were called following Ruelle and Takens, example "strange attractors". These sets can be characterized as following:

1) They have complicated topological structure. Unlike the attracting point, cycle and torus, which are the wellknown objects of the classical qualitative theory of ordinary differential equations, the "strange attractors" have, as a rule, a topological type of Cantor's set.

2) The behaviour of the orbits in the neighbourhood of "strange attractor" is in the highest degree chaotic. This can be a reflection of the certain stochastic properties inherenting to the dynamical system in the neighbourhood of attractor. The following property of them is, perhaps, most interesting not only for mathematicians but for physicists:

P1. There exists Borel measure μ which is invariant under dynamical system f, concentrated on attractor Λ, and such that for any continuous function h, defined in neighbourhood U of Λ

$$\frac{1}{n_k}\sum_{l=0}^{n_k-1}\int_U h(f^l(x))\,dm(x) \to \int_\Lambda h(x)\,d\mu(x) \qquad (1)$$

when $n_k \to \infty$. Here m is Lebesgue measure (Riemannian volume) in U and n_k is a sequence of integers.

The attractor is called stochastic if it satiesfies condition P1. According to an earlier definition given by Sinaĭ (cf. [15]) the attractor is stochastic if map $f|\Lambda$ satiesfies the condition (1) and mixes with respect to μ. The last requirement just allows us to exclude from consideration the "uninteresting" cases such as the attracting point, cycle and so on. The ergodic properties of map $f|\Lambda$ with respect to μ (such as ergodicity, mixing, positivity of the entropy, the K-property, the Bernoulli property and so on) and condition (1) are just the stochastic properties of dynamical system in neighbourhood of the attractor we are interested in.

3) The behaviour of the orbits on the very "strange attractor" is more or less unstable. It is this unstability that defines the certain stochastic properties of map $f|\Lambda$. This unstability can be expressed with help of more or less strong hyperbolicity conditions.

4. Attractor Λ is called hyperbolic if map $f|\Lambda$ satisfies the uniform and total hyperbolicity conditions. The stochastic properties of hyperbolic attractors were studied by Bowen and Ruelle (cf. [4], [3]) both for the discrete and continuous time. The requirement of the uniform and total hyperbolicity (as in the case of Anosov systems) guarantees the most powerful stochastic properties. For example, in the case of discrete time measure μ in (1) is uniquely defined and map $f|\Lambda$ is isomorphic with respect to μ to Bernoulli automorphism. On the other hand the hyperbolic attractors arise very rarely: Grines, Ẑuẑoma [9] and Plikin [12] have given a topological classification of a co-dimension one hyperbolic attractors recently.

Sinaĭ and the author stated recently the stochasticity of partially hyperbolic attractors (i.e. such that map $f|\Lambda$ satisfies the uniform and partial hyperbolicity conditions). Corresponding exact formulations are given in paragraph 2.

5. For the investigation of "strange attractor", arising in concrete physical models it is necessary to find such conditions which in the first place would be not difficult to verify and in the second place would open a possibility to state the stochasticity of attractor Λ. In this, the "verification" usually means some algorithmical procedure, realizing numerically with the help of computer. The uniform hyperbolic conditions (total or even partial) do not suit they are too difficult to be verified. Therefore, it is only natural to try to replace the uniform conditions by non-uniform ones. The late as wellknown (cf. for example [10]) can be formulated in the following way:

<u>P2</u>. There exists f-invariant Borel measure μ on Λ such that the set

$$\Lambda' = \{ x \in \Lambda : \chi^+(x, v) \neq 0 \ \text{for any} \ v \in T_x M \}$$

has μ-measure equal to 1 (in the non-uniform partial hyperbolic case it is necessary to change the words "for any" in the definition of the set Λ' by "for some" ones.)

Here $\chi^+(x,v)$ is the Ljapunov characteristic exponent of the vector v at the point x, defined by formula

$$\chi^+(x,v) = \overline{\lim_{n\to\infty}} \frac{1}{n} \ln \| df^n v \|.$$

As it was mentioned above the methods of calculation of the exponents with the help of computer is now worked out quite well. However, I do not know effective ways for verification of condition P2.

The usual situation, arising in the investigation of certain physical models is represented as following

1) It is possible to find an open subset U of the phase space such that $\overline{f(U)} \subset U$. In this case the set $\Lambda = \bigcap_{n\geq 0} f^n(U)$ is the closed f-invariant set (the attractor).

2) For almost any $x \in U$ (with respect to Riemannian volume m in U) and almost any vector $v \in T_x M$ Ljapunov characteristic exponent $\chi^+(x,v) > 0$ (and, as a rule, this value does not depend on x).

The condition 2) is the basis for the assumption that there is some unstability on set Λ. Does the condition of the non-uniform (total or partial) hyperbolicity and the stochasticity of the attractor follow from condition 2)? Apparently it is necessary to have some additional assumptions. The point is that the non-uniform hyperbolic conditions are very weakly connected with topology and, perhaps, we shall need "hard" requirements to state the stochasticity. For example, in the case when Λ is Lorentz attractor map f/Λ satisfies the conditions which are more strict then non-uniform hyperbolic ones. Nevertheless, these conditions can be verified with the help of a computer (cf. [10]) and they are sufficient to prove the stochasticity of this attractor (cf. [7]).

2. Partially Hyperbolic Attractor.

In this section it is formulated the results of the partial hyperbolic attractors, obtained recently by Sinaĭ and the author.

Let M be a smooth Riemannian manifold, $U \subset M$ be an open set, $f: U \to M$ be a $C^{1+\alpha}$ - diffeomorphism. Assume that

A1. There exists open subset $U_0 \subset U$ such that

$$\overline{f(U_0)} \subset U_0$$

The set

$$\Lambda = \bigcap_{n \geq 0} f^n(U_0)$$

is called the attractor. It is easily to see that Λ is closed and invariant under f and f^{-1}. It is also locally maximum set and

$$\Lambda = \bigcap_{n \geq 0} f^n(V_0)$$

for any neighbourhood $V_0 \subset U_0$ of Λ. Denote by $\Gamma^0(T\Lambda)$ the space of the continuous vector fields on Λ with uniform norm. Diffeomorphism f induces linear operator f_* in $\Gamma^0(T\Lambda)$, which action is given by the formule

$$(f_* v)(x) = df\, v\,(f^{-1}(x)),\ x \in \Lambda,\ v \in \Gamma^0(T\Lambda)$$

It is known that the spectrum of this operator consists of the whole circles. Attractor Λ is called partially hyperbolic if

A2. spectrum is contained in two rings which radii are $\lambda_1, \mu_1, \lambda_2, \mu_2$ respectively where $\infty > \lambda_1 \geq \mu_1 > \lambda_2 \geq \mu_2 > 0,\ \mu_1 > 1$.

Attractor Λ is called hyperbolic if in addition $\lambda_2 < 1$.

2. An example of a diffeomorphism with partially hyperbolic attractor is got if we take cartesion product of a diffeomorphism, having a hyperbolic attractor, on indentity map of some manifold. One can show that any small pertubation of this construction leads to a diffeomorphism having also a hyperbolic attractor, which is homeomorphic to unpertubated one.

One can get an other example, considering a family of the diffeomorphisms, obtained from cartesian product of two flows, each of which has a hyperbolic attractor. Any diffeomorphism of this family and also any its sufficiently small pertubation have a partial hyperbolic attractor.

3. It is known that the existence of smooth unstable manifold $W^u(x)$ for every $x \in \Lambda$ follows from partial hyperbolicity conditions. This manifold characterized the following property: for any $y \in W^u(x)$the orbits $\{f^n(y)\}$ and $\{f^n(x)\}$ draw together with uniform exponential rate when $n \to \infty$. It is easily to prove that

$$W^u(x) \subset \Lambda \qquad \text{for any } x \in \Lambda$$

Furthermore, $W^u(x)$ form continuous f-invariant C^1 - lamination on We shall assume that dim $W^u(x)$ = const.

4. We denote by $\mathcal{F}(\Lambda)$ the set of all functions φ which are continuous on Λ , have no nought and satisfy Cülder's condition. We define for $x \in \Lambda$, $y \in W^u(x)$ and integer $n \geq 0$ function

$$\rho_{\varphi,n}(x,y) = \prod_{k=0}^{n-1} [\varphi(f^{-k}(x))][\varphi(f^{-k}(y))]^{-1}$$

Proposition 1. 1) For any $x \in \Lambda$, $y \in W^u(x)$ there exists

$$\rho_\varphi(x,y) = \lim_{n\to\infty} \rho_{\varphi,n}(x,y) > 0.$$

2) Function $\rho_\varphi(x,y)$ is continuous.

3) For any $x \in \Lambda$, $y, z \in W^u(x)$

$$\rho_\varphi(x,y)\,\rho_\varphi(y,z) = \rho_\varphi(x,z)$$

We denote by $B^u(x,r)$ the open ball on manifold $W^u(x)$ with center at x and radius r. Let $x \in \Lambda$ and W be a local manifold, passing through x, having an additional dimension and being transversal to lamination W^u. The set

$$\Pi = \Pi(x, W, r) = \bigcup_{y \in W \cap \Lambda} B^u(y,r)$$

is called Λ-rectangle at point x. Let ν^u_y denote the measure in $W^u(y)$, induced by Riemannian metric and $\xi = \xi(\Pi)$ denote the partition of Π in the sets $B^u(y,r)$. This partition is measurable with respect to any Borel measure μ on Λ, since it is continuous. We denote by $\chi(y)$, $y \in \Pi(x,W,r)$ a point in $W \cap \Lambda$ such that $y \in B^u(\chi(y),r)$ (evidently this point is the only).

Let μ be a Borel measure on Λ. We fix $\varphi \in \mathcal{F}(\Lambda)$, $x \in \Lambda$ and Λ -rectangle $\Pi(x,W,r)$ and denote by $\mu_\xi(y)$ the conditional measures on elements $C_\xi(y) = B^u(\chi(y),r)$ of partition ξ. Measure μ is called Gibbs measure, if for any choise of point x and Λ -rectangle $\Pi(x,W,r)$ measures $\mu_\xi(y)$ satisfy the condition

$$d\mu_\xi(y) = r(\chi(y))\,\rho_\varphi(y,\chi(y))\,d\nu^u_y(y),$$

where $r(\chi(y))$ is a "normaled factor".

5. Let $\Phi(\Lambda)$ be a set of all f-invariant Gibbs (u,φ) -measures on Λ, $\varphi \in \mathcal{F}(\Lambda)$. We denote by $J^u_\mu(x)$, $\mu \in \Phi(\Lambda)$, $x \in \Lambda$ jacobian of map $df\,|\,TW^u(x)$ with respect to μ. Two following assertions describe some ergodic properties of $f|\Lambda$ with respect to $\mu \in \Phi(\Lambda)$. The first gives an estimation of the entropy from below, and the second can be used for the study of the K-property.

Theorem 1. Entropy $h_\mu(f|\Lambda)$ of automorphism $f|\Lambda$ with respect to $\mu \in \Phi(\Lambda)$ admits the following estimation:

$$h_\mu(f|\Lambda) \geq \int_\Lambda \ln J^u_\mu(x)\,d\mu(x) > 0$$

Theorem 2. There exists a set $\Lambda' \subset \Lambda$ and measurable partition ξ of Λ' such that

1) Λ' is f-invariant and open in Λ;

2) for any $x \in \Lambda'$ element $C_\xi(x)$ is open and connected subset of layer $W^u(x)$, where diam $C_\xi(x) \leq$ const (independent of x) and $\nu^u(\partial C_\xi(x)) = 0$

3) $f\xi \geq \xi$, $\bigvee_{n \geq 0} f^n \xi = \varepsilon$;

4) if $\mu \in \Phi(\Lambda)$ is ergodic measure then $\mu(\Lambda') = 1$ (in particular, Λ' is every where dense in Λ) and $\bigwedge_{n \geq 0} f^n \xi = \nu(W^u)$ where $\nu(W^u)$ is measurable cover of the partition of Λ in the layers of $W^u(x)$ with respect to μ.

6. Let m be a Riemannian volume in neighbourhood U of set Λ. It is known that $J^u_\mu \in \mathcal{F}(\Lambda)$. We shall call Gibbs measure simply u-measure.

Let $V_0 \subset U_0$ be a neighbourhood of Λ. We consider initial measure in V_0 given by non-negative density $\psi(x)$. We denote by $C(U_0)$ the Banach space of all continuous functions on V_0 with the uniform norm and define in this space the sequence of linear functionals by the formula

$$\ell_n(V_0,\psi)(h) = \frac{1}{n}\sum_{k=0}^{n-1} \int_{V_0} \psi(x)\,h(f^k(x))\,dm(x),$$

where $h \in C(\overline{V}_0)$. It is easily to see that family $\{\ell_n\}$ is relatively compact in the weak*topology. Let $\ell(V_0, \Psi)$ be some limit functional and $\mu = \mu(\ell, V_0, \Psi)$ a correponding Borel measure (according to Riss's theorem). Measure μ is, obviously, f-invariant and concentrated on set Λ.

Theorem 3. μ is u-measure.

Not only the existence of u-measures but the stochasticity of partial hyperbolic attractors follow from this theorem.

7. We describe the other way to construct n-measures on Λ. We fix $x \in \Lambda$ and choose a small open neighbourhood V_0 of x on $W^u(x)$. $\tilde{\nu}$ is a measure on U_0 given by the formula

$$d\tilde{\nu}(y) = \rho_{J^u_\nu}(x, y)\, d\nu^u_x(y)$$

Let ν_n is a sequence of the Borel measures such that for any Borel set $A \subset \Lambda$

$$\nu_0(A) = \tilde{\nu}(A \cap U_0), \quad \nu_n(A) = \nu_0(f^{-n}(A))$$

Theorem 4. Any limit point of a sequence $\mu_n = \frac{1}{n}\sum_{k=0}^{n-1} \nu_k$ is f-invariant u-measure.

References

[1] Anosov D.V. Geodesic Flows on Closed Riemannian Manifolds of Negative Curvature, Trudy Mat Inst. Steklov 90 (1967); English Transl., Proc.Steklov Inst.Math. 90 (1967).

[2] Anosov D.V., Ya.G.Sinaĭ. Certain Smooth Ergodic Systems, Uspehi Mat. Nauk 22, no.5 (1967) 107-172; English Transl.in Russian Math.Surveys 22 (1967).

[3] Bowen R. Equilibrium States and the Ergodic Theory of Anosov Diffeomorphisms, Lect.Botes 470 (1975), 108.

[4] Bowen R., Ruelle D. The Ergodic Theory of Axiom A Flows, Invent. Math., 29 (1975), 181-202.

[5] Bunimovich L.A. On Billiards Close to Dispersing, Mat.Sb. 94 (1974) 49-73, Math.USSR Sb., 23 (1974), 45-67.

[6] Bunimovich L.A., Sinaĭ Ya.G. On a Fundamental Theorem in the Theory of Dispersing Billiard, Mat.Sb. 90 (1973) 415-431; Math.USSR Sb. 19 (1973), 407-424.

[7] Bunimovich L.A., Sinaĭ Ya.G. Stochasticity of the Attractor in the Lorentz model, in Nonlinear Wave, Proc.of Winter School, Nauka Press, 1979.

[8] Bunimovich L.A., Sinaĭ Ya.G. Markov Partition for Dispersed Billiards, Comm.Math.Phys. 77 (1980), 247-280.

[9] Grines V.Z., Žužoma E.V. On Topological Classification of Orientable Attractor over n-dimensional Torus, Uspehi Mat.Nauk, 34 N 4 (1979)

[10] Pesin Ya.B. Characteristic Ljapunov Exponents and Smooth Ergodic Theory, Uspehi Mat.Nauk, 32, N 4 (1977) 55-112; Russian Math. Surveys 32 (1977).

[11] Pesin Ya.B., Sinaĭ Ya.G. The Hyperbolicity and Stochasticity of Dynamical System, Russian Math.Surveys, Benjamin Press, 1970.

[12] Plikin R.V. On Topological Classification of C-dimensional I Hyperbolic Attractor, Uspehi Mat.Nauk, 35,N 3-4 (1980).

[13] Ruelle D., Takens F. On the Nature of Turbulence, Comm.Math.Phys. 90 (1971),167-192.

[14] Sinaĭ Ya.G. Classical Systems With Countable Lebesgue Spectrum II, Izv.Akad.Nauk SSSR, Ser.Mat. 30 (1966), 15-68, Amer.Math.Soc.Trans. (2) 68 (1968), 34-88.

[15] Sinaĭ Ya.G. Stochasticity of Dynamical System, in Non-linear Waves, Proc.of Winter School, Nauka Press (1979).

[16] Sinaĭ Ya.G., Vul E.B. Hyperbolicity Conditions for the Lorentz Model, Phisica 2D (1981),3-7.

TWO NOTES ON THE INDIVIDUAL ERGODIC THEOREM.

Beloslav Riečan [1)]

The paper consists of two parts. The first part is devoted to the non-commutative probability theory, the second part to the non-additive probability theory.

1

Instead of a probability space (X,S,P) one can study an abstract logic L with a probability measure m: $L \longrightarrow R$ (so called state, see [7]). These notions seem to be adequate to the quantum mechanical systems and they are defined as follows.

Logic L is a σ-complete lattice with the first and the last element 0 and 1, respectively, with an orthocomplementation $\perp: a \longrightarrow a^{\perp}$, which satisfies

(i) $(a^{\perp})^{\perp} = a$ for all $a \in L$;

(ii) if $a \leqq b$, then $b^{\perp} \leqq a^{\perp}$;

(iii) $a \vee a^{\perp} = 1$ for all $a \in L$;

and the orthomodular law holds in L: if $a \leqq b$, then $b = a \vee (b \wedge a^{\perp})$. (The family of all linear subspaces of a Hilbert space presents a very instructive and important example of a logic.)

We say that two elements $a,b \in L$ are orthogonal ($a \perp b$), if $a \leqq b^{\perp}$. A state is a map m: $L \longrightarrow \langle 0,1 \rangle$ such that

(i) $m(1) = 1$;

(ii) $m(\vee a_i) = \sum m(a_i)$ for any sequence $(a_i)_i$ of pairwise orthogonal elements of L.

Instead of random variables f: $X \longrightarrow R$ one can study so-called observables x: $B(R) \longrightarrow L$, where B(R) is the family of all Borel subsets of the reals R. (In the classical (X,S,P)-case we can define $x(E) = f^{-1}(E) \in S$ for $E \in B(R)$.) An observable is a map x: $B(R) \longrightarrow L$ such that

(i) $x(\emptyset) = 0$;

(ii) if $E \cap F = \emptyset$, then $x(E) \perp x(F)$;

(iii) $x(\cup E_i) = \vee x(E_i)$ for any sequence $(E_i)_i$ of pairwise disjoint Borel sets.

Finally instead of an endomorphism of a probability space (X,S,P) we shall investigate a homomorphism of a logic L, i.e. a mapping τ : $L \longrightarrow L$ satisfying the following conditions:

---

1) Department of Probability and Statistics, Comenius University, Bratislava, Czechoslovakia.

(i) $\tau(0) = 0$;

(ii) $\tau(a^{\perp}) = \tau(a)^{\perp}$ for all $a \in L$;

(iii) $\tau(\bigvee a_i) = \bigvee \tau(a_i)$ for every sequence $(a_i)_i$ of elements of L.

A homomorphism $\tau: L \longrightarrow L$ is m-preserving (where $m: L \longrightarrow R$ is a state), if $m(\tau(a)) = m(a)$ for all $a \in L$.

The main tool of our investigations is the representation of homomorphisms of a logic by point Borel measurable transformations of R.

<u>Theorem 1.</u>([3] , Prop. 1) Let $x: B(R) \longrightarrow L$ be an observable, let $\tau: L \longrightarrow L$ be a homomorphism. Then the following two statements are equivalent:

(i) There is a Borel measurable transformation $T: R \longrightarrow R$ such that $\tau(x(E)) = x(T^{-1}(E))$ for every $E \in B(R)$.

(ii) τ is x-measurable, i.e. $\tau(x(B(R))) \subset x(B(R))$.

Since all the observables $\tau^i \circ x$ (i=0,1,2,...) map the Borel σ-algebra B(R) into the σ-algebra x(B(R)), there exist the sums $\sum_{i=0}^{n-1} \tau^i \circ x$ for all n ($\tau^i \circ x$ are compatible, see [7]). Namely then there exists a homomorphism $h: B(R^n) \longrightarrow L$ such that $\tau^i \circ x = h \circ p_i^{-1}$, where $p_i: R^n \longrightarrow R$ is the projection of R^n into the i-th coordinate. The sum $\sum_{i=0}^{n-1} \tau^i \circ x = y: B(R) \rightarrow L$ is defined by the formula $y(E) = h(g^{-1}(E))$, where $g: R^n \longrightarrow R$, $(t_1, \dots, t_n) \longrightarrow t_1 + \dots + t_n$.

For the presentation of the main result we need the following three definitions. If $x: B(R) \longrightarrow L$ is an observable, then $y = \frac{1}{n}x: B(R) \rightarrow L$ is an observable defined by $y(E) = x(\{nt;\ t \in E\})$, $E \in B(R)$.

A sequence $(x_n)_n$ of observables converges m-a.e. to an observable $x: B(R) \longrightarrow L$, if $m(\inf_k \sup_{n \geq k} (x_n - x)(\langle -\varepsilon, \varepsilon \rangle^c)) = 0$ for every $\varepsilon > 0$.

Finally define $m_x: B(R) \rightarrow R$ by $m_x(E) = m(x(E))$ and put $m(x) = \int_R t \, dm_x(t)$ if the integral exists. (Similarly as in the classical case, m(x) presents the mean value of x.)

<u>Theorem 2.</u> Let x be an observable, m be a state, τ be an x-measurable m-preserving homomorphism of a logic L, i.e. $\tau(x(B(R))) \subset x(B(R))$. If there exists m(x), then there exists a Borel measurable function $s: R \rightarrow R$ such that $\frac{1}{n} \sum_{i=0}^{n-1} \tau^i \circ x$ converges m-a.e. to $x \circ s^{-1}$.

Theorem 2 has been proved in [3] but only for ergodic homomorphisms τ with $m(x) = 0$. In the general form it will appear in a paper by P. Vrábel (see [8]). It presents a definitive result achieved by the representation method. Of course, there would be convenient to obtain more general results using some other methods.

2

The idea to develop a subadditive probability theory is due to L. Breiman [1] . Of course, by our knowledge there is no remarkable result in this direction. On the other hand there are many interesting results concerning subadditive measure and integration theory.

By a submeasure we shall call (in this article) a mapping $m: S \longrightarrow R$ satisfying the following conditions:

(i) S is a σ-algebra of subsets of a set X;

(ii) $m(A) \leq m(B)$ for every $A, B \in S$ such that $A \subset B$;

(iii) $m(A \cup B) + m(A \cap B) \leq m(A) + m(B)$ for every $A, B \in S$;

(iv) $m(A_n) \searrow 0$ for every non-increasing sequence $(A_n)_n$ of sets from S with empty intersection.

If f is a non-negative measurable function defined on X and $D = \{a_0, a_1, \ldots, a_n\}$ is a collection of non-negative real numbers such that $0 = a_0 < a_1 < \ldots < a_n$, then we put

$$S(f,D) = \sum_{i=1}^{n} (a_i - a_{i-1}) \cdot m(\{t ; f(t) \geq a_i\}).$$ The integral $\int f\, dm$ can be defined by the formula $\int f\, dm = \sup S(f,D)$, where the supremum is taken over all finite collections $D: 0 = a_0 < a_1 < \ldots < a_n$ (see [6]).

All what we are able to prove is the maximal ergodic theorem. Of course, since the integral need not be additive, the individual ergodic theorem is not an immediate consequence of the maximal one and the validity of the individual ergodic theorem in non-additive case is open.

Theorem 3. Let m be a submeasure defined on a σ-algebra S of subsets of a set X. Let f be a non-negative measurable function, $a \in R$, $a \geq 0$. Let $T: X \longrightarrow X$ be an m-preserving mapping, i.e. $m(T^{-1}(E)) = m(E)$ for every $E \in S$. Put $A = \{x \in X ; \exists k \in N,\ f(x) + f(T(x)) + \ldots + f(T^k(x)) > ka\}$. Then $\int f \chi_A\, dm \geq a\, m(A)$.

References

[1] Breiman, L.: Probability theory. Reading 1968.

[2] Dvurečenskij, A.: Law of large numbers and the central limit theorems on a logic. Math. Slovaca 29 (1979), 397 - 410.

[3] Dvurečenskij, A. - Riečan, B.: On the individual ergodic theorem on a logic. Comment. Math. Univ. Carol. 21, 2 (1980), 385 - 391.

[4] Harman, B.: On the maximal ergodic theorem, to appear.

[5] Pulmannová, S.: Individual ergodic theorem on a logic, to appear.

[6] Šipoš, J.: Integral with respect to a pre-measure. Math. Slovaca 29 (1979), 141 - 155.

[7] Varadarajan, V.S.: Geometry of Quantum Theory. New York 1968.

[8] Vrábel, P.: On the individual ergodic theorem on a logic, to appear.

TWO APPLICATIONS OF A THEOREM OF GLIMM AND EFFROS IN ERGODIC THEORY

Klaus Schmidt

§1. The Theorem of Glimm and Effros

Let G be a locally compact second countable group, X a complete separable metric space, and let $(g,x) \to gx$ be a continuous action of G on X (this means that $(g,x) \to gx$ is a continuous map from $G \times X$ to X satisfying $ex = x$ for all $x \in X$, where e is the identity element of G, and $g_1(g_2x) = (g_1g_2)x$ for all $g_1, g_2 \in G$, $x \in X$). S will denote the Borel field of X. A probability measure μ on (X,S) is said to be *quasi-invariant* under G if, for every $g \in G$ and $B \in S$, $\mu(gB) = 0$ if and only if $\mu(B) = 0$. A quasi-invariant probability measure μ on (X,S) is called *properly ergodic* under G if every G-invariant set $B \in S$ satisfies $\mu(B) \cdot (1-\mu(B)) = 0$, and if $\mu(Gx) = 0$ for every $x \in X$, where $Gx = \{gx : g \in G\}$.

It is easy to find actions of groups without properly ergodic measures: if G acts on itself or on any of its homogeneous spaces by translation there can be no properly ergodic measures. The same is true if X is a 'nice' union of homogeneous G-spaces in the sense that there exists a Borel cross-section of the orbits, i.e. a Borel set $B \subset X$ which intersects every G-orbit in exactly one point. The converse is not at all trivial: in 1961 J. Glimm proved that G has properly ergodic quasi-invariant probability measures on (X,S) if and only if there exists no such set (cf. [6]). Glimm used the additional assumption that X is locally compact, but this condition was later removed by E. Effros in [3]. A different, and, as it turns out, very useful criterion for the existence of properly ergodic measures on (X,S) in [3,6] is formulated in terms of the topology of individual G-orbits. For every $x \in X$, let $G_x = \{g \in G : gx = x\}$ denote the stability subgroup of x. The map $gG_x \to gx$ from the space G/G_x of left cosets of G_x to the orbit Gx of x under G is clearly a bijection. We can now state the promised theorem of Glimm and Effros:

1.1. Theorem. The following conditions are equivalent.

(1) G has a properly ergodic quasi-invariant probability measure on (X,S).

(2) There exists a point $x_0 \in X$ for which the map $gG_{x_0} \to gx_0$ from G/G_{x_0} to Gx_0 is not a homeomorphism onto its image (in the subspace topology).

Note that condition (2) is equivalent to the existence of a sequence $(g_n : n \geq 1)$ in G with

(3) $\lim_n g_n G_{x_0} = \infty$ in G/G_{x_0}, but

(4) $\lim_n g_n x_0 = x_0$ in X (note that we may assume that $g_n x_0 \neq x_0$ for every $n \geq 1$).

---

1) Mathematics Institute, University of Warwick, Coventry CV4 7AL

This suggests to call the point x_0 *recurrent*. Theorem 1.1 can now be expressed as follows: the action of G on X has a recurrent point if and only if G has a non-singular, properly ergodic probability measure μ on (X,S). Note that, if G is discrete, then a point $x_0 \in X$ is recurrent if and only if every deleted neighbourhood of x_0 contains a point of Gx_0. For $G = Z$ this theorem of Glimm and Effros was re-proved and strengthened in Katznelson-Weiss [8] and Kornfeld [9] to show that the existence of a recurrent point is equivalent to the existence of uncountably many mutually singular quasi-invariant and properly ergodic measures μ for $G = Z$ on (X,S). In particular, μ may be chosen to be of any given type in the sense of Krieger's classification (cf. [9, 10, 11]), with the possible exception of type II_1 (G may not have any invariant probability measures on (X,S)).

The basic idea in the proof of Theorem 1.1 (and of the related results in [8, 9]) is quite simple. General recurrence considerations for properly ergodic group actions imply that, for every quasi-invariant and properly ergodic probability measure μ for G on (X,S), μ-a.e. $x \in X$ is recurrent under G.

For the converse I shall outline a proof modelled on the argument in [8]. Let $x_0 \in X$ be a recurrent point for G. Using the G-orbit of x_0 we shall construct an adding machine (odometer) on which 'part' of G acts, use one of the many quasi-invariant and properly ergodic measures of the adding machine to obtain a measure on Y which is properly ergodic under this 'part' of G, and finally 'smear' the measure using the Haar measure of G. The resulting measure will be quasi-invariant and properly ergodic under the action of G on (X,S).

To make this idea more precise we choose and fix a bounded metric d on X and put, for every $\varepsilon > 0$, $B(\varepsilon) = \{x \in X: d(x,x_0) \le \varepsilon\}$. The diameter of a set $A \subset X$ will be denoted by dia (A). Finally we write $\mathbf{Z}_2 = \mathbf{Z}/2\mathbf{Z} = \{0,1\}$ and denote by $\mathbf{Z}_2^n$ and $K = \mathbf{Z}_2^{\mathbf{N}} = \{(k_1,k_2,..): k_i \in \mathbf{Z}_2$ for all $i\}$ the Cartesian products of n and of countably many copies of $\mathbf{Z}_2$, respectively. K is a compact metric abelian group in its product topology and under co-ordinate-wise addition (mod 2), and we denote its Borel field by $\mathcal{K}$ and its (normalised) Haar measure by λ. $\Gamma \subset K$ will denote the countable subgroup $\{k \in K: k = (k_1,k_2,\ldots)$ with $k_n = 1$ for only finitely many $n \ge 1\}$. Γ acts on $(K, \mathcal{K}, \lambda)$ by translation, and this action is measure preserving and ergodic.

We now write G_{x_0} for the stability subgroup of the recurrent point $x_0 \in X$ and choose sequences $(g_n: n \ge 1)$ in G and $(\delta_n: n \ge 1)$ of positive real numbers such that the following conditions are satisfied.

(a) for every compact set $C \subset G/G_{x_0}$ there exists an $N \ge 1$ with $g_n\, g_{n-1}^{k_{n-1}} \cdots g_1^{k_1} G_{x_0} \not\subset C$ for every $n > N$ and every $(k_1,\ldots,k_{n-1}) \in \mathbf{Z}_2^{n-1}$,

(b) $\lim_n g_n x_0 = x_0$,

(c) $\lim_n \sup_{(k_1,\ldots,k_n)\in\mathbf{Z}_2^n} \operatorname{diam} (g_n^{k_n} \cdots g_1^{k_1} B(\delta_n)) = 0$,

d) for every $n \geq 1$, $g_n B(\delta_n) \cap B(\delta_n) = \phi$ and $g_{n+1}B(\delta_{n+1}) \cup B(\delta_{n+1}) \subset B(\delta_n)$.

For every $k = (k_1,k_2,\ldots) \in K$, consider the intersection $\bigcap_{n\geq 1} g_n^{k_n} \ldots g_1^{k_1} B(\delta_n) \subset X$.

Conditions (c) and (d) imply that this intersection contains at most one point, and the completeness of X implies that this intersection is non-empty. We can thus define a Borel map $\phi: K \to X$ with $\{\phi(k)\} = \bigcap_{n\geq 1} g_n^{k_n} \ldots g_1^{k_1} B(\delta_n)$ for every $k \in K$, and (c) and (d) show that ϕ is injective and continuous. Let G_0 be the countable subgroup of G generated by $\{g_n : n \geq 1\}$. From our construction it is clear that we can find, for every $k \in K$ and $\gamma \in \Gamma$, an element $g(\gamma,k) \in G_0$ with $\phi(\gamma + k) = g(\gamma,k)\phi(k)$. Put $\nu = \lambda\phi^{-1}$, choose a strictly positive function $\psi: G \to \mathbb{R}$ with $\int \psi(g)dg = 1$, where integration is taken with respect to the left Haar measure on G, and define a new measure μ on (X,S) by $\mu(E) = \int \nu(gE)\psi(g)dg$, $E \in S$. Clearly μ is quasi-invariant and ergodic under the action of G on (X,S), and it is not difficult to verify that μ is properly ergodic, i.e. not concentrated on a single G-orbit (the last assertion turns out to be a consequence of conditions (a) and (b) in our choice of the sequence $(g_n : n \geq 1)$). This completes the outline of the proof of Theorem 1.1.

§2. Spectra of ergodic group actions

Let G be a locally compact, second countable, non-compact abelian group, (X,μ) a σ-finite Lebesgue space, and $(g,x) \to gx$ a non-singular, properly ergodic action of G on (X,μ). An element $\gamma \in \Gamma = \hat{G}$ (the character group of G) is said to belong to the L^∞-spectrum Sp(G,X) of the action of G on (X,μ) if there exists a non-zero function $f \in L^\infty(X,\mu)$ with $f(gx) = \gamma(g)f(x)\mu$ - a.e., for every $g \in G$. Sp(G,X) is clearly a subgroup of Γ, and it is not difficult to verify that it is also a Borel set. If the action of G is finite measure preserving, Sp(G,X) is necessarily countable, but there exist examples of $\mathbb{Z}$-and $\mathbb{R}$-actions with uncountable spectrum. It has also been known for some time that Sp(G,X) cannot be equal to Γ, but there are many open problems concerning the possible size and form of spectra even in the case $G = \mathbb{Z}$. Using the theorem of Glimm and Effros in Section 1 we can give a complete *metric* characterization of the possible subgroups which occur as Sp(G,X), although it is still not possible to give an explicit description of these groups (except for some examples, of course).

2.1. Theorem Let G be a locally compact, second countable, non-compact abelian group with character group Γ, and let σ be a probability measure on Γ with Fourier transform $\hat{\sigma}$. The following conditions are equivalent.

(1) There exists a non-singular, properly ergodic action $(g,x) \to gx$ of G on (X,μ) such that $\sigma(\mathrm{Sp}(G,X)) = 1$,

(2) $\limsup_{g\to\infty} \mathrm{Re}\, \hat{\sigma}(g) = 1$, where Re denotes the real part.

Outline of proof: Let $U(\Gamma,\sigma,S^1)$ denote the group of σ-equivalence classes of Borel functions on Γ of modulus 1 under pointwise multiplication and in the topology of convergence in σ-measure. $X = U(\Gamma,\sigma,S^1)$ is a complete, separable, metric group, and

G acts on X continuously by $(g,x) \to gx$ with $(gx)(\gamma) = \gamma(g)x(\gamma)$ for every $g \in G$, $x \in X$, $\gamma \in \Gamma$. Condition (2) in Theorem 2.1 is equivalent to saying that the function 1 in X is a recurrent point for this action. Hence, if (2) is satisfied, we can apply Theorem 1.1 to obtain a non-singular and properly ergodic probability measure μ for the action of G on X. Consider now the function $F: \Gamma \times X \to S^1$, given by $F(\gamma,x) = x(\gamma)$. It is not difficult to show that F is well-defined $\sigma \times \mu$-a.e. and measurable, and Fubini's theorem shows that there exists a set $B \subset \Gamma$ with $\sigma(B) = 1$ such that, for every $\gamma \in B$, $g \in G$, $F(\gamma,gx) = \gamma(g)\, F(\gamma,x)$ μ-a.e.. This shows that $Sp(G,X) \supset B$ and hence that $\sigma(Sp(G,X)) = 1$. This proof is taken from [13].

The converse is proved in [12] and is not related to Theorem 1.1.

2.2. Remark. The G-action in Theorem 2.1 may be chosen to preserve an infinite, σ-finite measure equivalent to μ.

2.3. Remark. Let G be a locally compact, non-compact, second countable group, and let $(g,x) \to gx$ be a properly ergodic non-singular action of G on a Lebesgue space (X,μ). As in the case where G is abelian we shall say that a non-zero function $f \in L^\infty(X,\mu)$ is an *eigenfunction* of G if, for every $g \in G$, there exists a constant $\gamma(g) \in \mathbb{C}$ with $f(gx) = \gamma(g)f(x)$ μ-a.e. Clearly $\gamma: G \to \mathbb{C}$ is a continuous homomorphism, and $\gamma(g) \in S^1$ for every $g \in G$. Put Γ equal to the group of continuous homomorphisms $\gamma: G \to S^1$ ($\Gamma = G/[G,G]$, where $[G,G]$ denotes the closed commutator subgroup of G). We define again $Sp(G,X) \subset \Gamma$ as the set of all γ for which there exists an eigenfunction $f \in L^\infty(X,\mu)$ with $f \circ g = \gamma(g)f$ for every $g \in G$. If $[G,G]$ is non-compact, one can construct actions of G on (X,μ) which are non-singular and properly ergodic but for which $Sp(G,X) = \hat{G}$.

We now turn to an application of Theorem 2.1 which describes the number of ergodic elements in an ergodic flow.

2.4. Corollary. Let σ be a probability measure on $\mathbb{R}$ with $\sigma(\{0\}) = 0$. The following conditions are equivalent.

(1) There exists a flow $(T_t : t \in \mathbb{R})$ on a Lebesgue space (X,μ) which is non-singular and properly ergodic and satisfies $\sigma\{s \in \mathbb{R} : s \neq 0 \text{ and } T_{1/s} \text{ is non-ergodic}\} = 1$.

(2) $\limsup_{s\to\infty} \operatorname{Re} \hat{\sigma}(s) = 1$.

This is proved in [12] and [13]. The connection between Corollary 2.3 and Theorem 2.1 comes from the fact that $T_{1/s}$ is non-ergodic if and only if $ks \in Sp((T_t),X)$ for some $k \geq 1$. An obvious consequence of Corollary 2.3 is the fact that ergodic flows have ergodic elements and that the set $\{s : T_{1/s} \text{ is non-ergodic}\}$ must have Lebesgue measure zero, but may be uncountable.

§3. Mild mixing

Let (X,μ) and (Y,ν) be Lebesgue probability spaces and let T be a measure preserving, ergodic automorphism of (X,μ). As is well known, T is weakly mixing if and only if, for every measure preserving, ergodic automorphism V of (Y,ν), $T \times V$ is again ergodic on $(X \times Y, \mu \times \nu)$. One may now ask whether the same is still true if V is only assumed to be non-singular and properly ergodic on (Y,ν). This question was studied by Furstenberg and Weiss in [5] and answered negatively. They were thus led to introduce the notion of mild mixing: T is called *mildly mixing* if $T \times V$ is ergodic for every ergodic non-singular automorphism V of (Y,ν). The main result in [5] states that mild mixing is - like weak mixing - a spectral property: T is mildly mixing if and only if $\liminf_{n\to\infty} \mu(T^n B \,\Delta\, B) > 0$ for every measurable set $B \subset X$ with $0 < \mu(B) < 1$. Although their proof was constructive, this result was later reproved by E. Flytzanis [4] using an argument related to Theorem 1.1. In order to explain this connection more clearly and in full generality we consider a non-singular, properly ergodic action $(g,x) \to gx$ of a locally compact, second countable group G on the Lebesgue probability space (X,μ). Following [5] we call this action mildly mixing if, for every properly ergodic non-singular action $(g,y) \to gy$ of G on (Y,ν), the product action $(g,(x,y)) \to (gx,gy)$ of G on $(X \times Y,\mu \times \nu)$ is again ergodic.

3.1. Theorem. The action $(g,x) \to gx$ of G on (X,μ) is mildly mixing if and only if $\liminf_{g\to\infty} \mu(gB \,\Delta\, B) > 0$ for every measurable set $B \subset X$ with $0 < \mu(B) < 1$.

Outline of proof: If the G-action is not mildly mixing, choose a non-singular, properly ergodic action $(g,y) \to gy$ of G on (Y,ν) such that the product action is non-ergodic. Let $D \subset X \times Y$ be a non-trivial measurable G-invariant set and put, for every $y \in Y$, $D_y = \{x \in X: (x,y) \in D\}$. The general recurrence properties of ergodic actions now imply the existence of a $y_0 \in Y$ with $0 < \mu(D_{y_0}) < 1$ and $\liminf_{g\to\infty} \mu(D_{y_0} \,\Delta\, gD_{y_0}) = 0$.

To prove the converse we let, as before, $U(X,\mu,S^1)$ denote the space of μ-equivalence classes of complex measurable functions on X of modulus 1 under convergence in μ-measure. Let $(g,\phi) \to g\phi$ denote the action of G on $Y = U(X,\mu,S^1)$ given by $(g\phi)(x) = \phi(g^{-1}x)$, and observe that this action is continuous. Assume now the existence of a set $B \subset X$ satisfying $0 < \mu(B) < 1$ and $\liminf_{g\to\infty} \mu(B \,\Delta\, gB) = 0$. In order to keep the proof simple we assume (*with* loss of generality) that $\mu(B \,\Delta\, gB) > 0$ whenever $g \neq e$ (the general case is proved in [15]). Put $\phi_0(x) = 2\chi_B(x) - 1$, where χ_B is the indicator function of B. From the choice of B it is clear that ϕ_0 is a recurrent point for the action of G on $Y = U(X,\mu,S^1)$, and Theorem 1.1 implies the existence of a probability measure ν on Y such that the action $(g,\phi) \to g\phi$ of G on (Y,ν) is non-singular and properly ergodic. Define $F: X \times Y \to S^1$ by $F(x,\phi) = \phi(x)$ and observe that F is well defined $\mu \times \nu$- a.e., measurable, and invariant under the production action of G on $(X \times Y, \mu \times \nu)$. Since F is non-constant, the product action is non-ergodic.

3.2. Remark. The assumption that the action of G on (X,μ) be non-singular (i.e. not necessarily measure preserving) is not as general as it may appear. If the action $(g,x) \to gx$ of G on (X,μ) is mildly mixing then there exists a probability measure $\dot{\mu}$ on X which is equivalent to μ and invariant under G. This somewhat surprising result was first proved in [1] for $G = \mathbb{Z}$, and the general result is contained in [15].

3.3. Remark. If the action of G on (X,μ) is not mildly mixing we can find an infinite Lebesgue space (Y,ν) and a properly ergodic, measure preserving action $(g,y) \to gy$ of G on (Y,ν) such that the product action of G on $(X \times Y, \mu \times \nu)$ is not ergodic (cf. [15])

In spite of their abstractness, Theorems 2.1 and 3.1 give additional insight even in the cases $G = \mathbb{Z}$ and $G = \mathbb{R}$. By exhibiting these results as consequences of Theorem 1.1 the proofs become not only more general, but also conceptually simpler than the corresponding ones for $G = \mathbb{Z}$ (I should point out here that B. Weiss has an unpublished proof of Theorem 2.1 for $G = \mathbb{Z}$ by a completely different method).

§4. Weak mixing and mild mixing

In this last section we discuss some problems related to Theorem 3.1 and to the connection between weak and mild mixing. Recall that a measure preserving, properly ergodic action $(g,x) \to gx$ of a locally compact second countable group G on a Lebesgue probability space (X,μ) is called *weakly mixing* if the product action $(g,(x_1,x_2)) \to (gx_1,gx_2)$ of G on $(X \times X, \mu \times \mu)$ is again ergodic (cf. [2]). This condition is equivalent to the assumption that, for every measure preserving, properly ergodic action $(g,y) \to gy$ of G on a Lebesgue probability space, the product action $(g,(x,y)) \to (gx,gy)$ is again ergodic. H.A. Dye proved in [2] that the action $(g,x) \to gx$ of G on (X,μ) is weakly mixing if and only if the unitary representation $(U_g f)(x) = f(g^{-1}x)$ of G on $H = L^2(X,\mu) \ominus \mathbb{C} = \{f \in L^2(X,\mu): \int f d\mu = 0\}$ has no finite dimensional subrepresentations. Theorem 3.1, on the other hand, is easily seen to imply that the action of G on (X,μ) is mildly mixing if and only if, for every $f \neq 0$ in H, $\limsup_{g\to\infty} \langle U_g f,f\rangle < \|f\|^2$, where $\langle .,.\rangle$ denotes the inner product and $\|.\|$ the norm in H. Clearly every mildly mixing action of G is weakly mixing. The converse is, of course, not true. One can find, for every abelian, or, more generally, solvable, non-compact, locally compact second countable group G a measure preserving properly ergodic action on a Lebesgue probability space which is weakly mixing, but not mildly mixing. The following two results are proved in [2,14] and provide the basis for further interesting problems. Let G be a locally compact second countable group.

4.1. Theorem. The following conditions are equivalent.

(1) Every finite measure preserving properly ergodic action of G is weakly mixing.

(2) G is minimally almost periodic.

4.2. Theorem. The following conditions are equivalent

(1) Every finite measure preserving properly ergodic action of G is mildly mixing,

(2) For every continuous, irreducible unitary representation V of G on a complex separable Hilbert space K, and for every $v \in K$, $v \neq 0$, we have $\limsup_{g\to\infty} \langle V_g v,v\rangle < \|v\|^2$.

In view of Theorem 4.2 we shall say that G has *property M* if it satisfies the conditions of Theorem 4.2. A result by R. Howe and C.C. Moore in [7] implies that every non-compact simple Lie group with trivial centre has property M. Using this result one can show that a non-compact *connected* locally compact second countable group G has property M if and only if it is a perfect compact central extension of a non-compact centre-less simple Lie group (cf. [14]). The results in [14] also allow one to give examples of totally disconnected groups with property M. There are, however, a number of open problems:

(1) Describe all zero-dimensional groups with property M.

(2) Are there amenable groups with property M?

(3) Are there discrete groups with property M?

Problem (3) seems particularly intriguing. In [14] it is proved that no infinite countable subgroup of an algebraic group over a commutative field has property M. In particular, the classical groups SL(n,Z), SL(n,Q), etc, all do not have properties M. If a countable group G has property M, it also has the following remarkable properties: (a) G is a finite central extension of a simple group G_0, (b) G is finitely generated, and (c) every proper subgroup of G is finite.

Groups with these properties (a) - (c) are known to exist, but I do not know whether any of them have property M.

References

[1] Aaronson, J., Lin, M., and Weiss, B.: Mixing properties of Markov operators and ergodic transformation, and ergodicity of Cartesian products. Israel J. Math. 33 (1979), 198-224.

[2] Dye, H.A., On the ergodic mixing theorem. Trans. Amer. Math. Soc. 118 (1965), 123-130.

[3] Effros, E.G., Transformation groups and C*-algebras. Ann. Math. 81 (1965), 38-55.

[4] Flytzanis, E., Rigid factors of ergodic transformations, Israel J. Math. 35 (1980), 323-326.

[5] Furstenberg, H., and Weiss, B., The finite multipliers of infinite ergodic transformations. Lecture Notes in Mathematics, Vol. 668, pp. 127-132. Springer 1978.

[6] Glimm, J., Locally compact transformation groups, Trans. Amer. Math. Soc. 101 (1961), 124-138.

[7] Howe, R., and Moore, C.C., Asymptotic properties of unitary representations. J. Functional Analysis 32 (1979), 72-96.

[8] Katznelson, Y., and Weiss, B., The construction of quasi-invariant measures. Israel J. Math. 12 (1972), 1-4.

[9] Kornfeld, I.P., Quasi-invariant measures for topological dynamical systems (Russian). Izv. Akad. Nauk. SSSR. Ser. Mat. 38 (1974), 1305-1323.

[10] Krieger, W., On the Araki-Woods asymptotic ratio set and non-singular transformations of a measure space. Lecture Notes in Mathematics Vol. 160, pp. 158-177. Springer 1970.

[11] Krieger, W., On Borel automorphisms and their quasi-invariant measures. Math. Z. 151 (1976), 19-24.

[12] Losert, V., and Schmidt, K., A class of probability measures on groups arising from some problems in ergodic theory. Lecture Notes in Mathematics, Vol. 706 pp. 220-238, Springer 1979.

[13] Schmidt, K., Spectra of ergodic group actions. To appear in Israel J. Math.

[14] Schmidt, K., Mildly mixing actions of locally compact groups II. Preprint 1981.

[15] Schmidt, K., and Walters, P., Mildly mixing actions of locally compact groups I. To appear in J. London Math. Soc.

CODES IN ERGODIC THEORY AND INFORMATION: SOME EXAMPLES

Štefan Šujan [1]

The aim of this paper is to point out several examples of interplay between ergodic theory and information theory. The examples include noiseless source coding using stationary codes as well as using block codes, and source coding with a fidelity criterion.

1. Noiseless Source Coding and Krieger's Theorem

Let E(B) denote the space of all ergodic sources over a discrete alphabet B. Let, for $[B,\lambda]$ and $[B,\tau] \in E(B)$,

$$d_w(\lambda,\tau) = \Sigma_{n=1}^{\infty} \Sigma_{y \in B^n} |\lambda^n(y) - \tau^n(y)| \;;$$

that is, d_w is a metric compatible with the topology of weak convergence in E(B).

Theorem 1. Let $[A,\mu,X]$ be an aperiodic ergodic source over a countable alphabet A having finite entropy H(X). Let $[B,\tau,U]$ be any ergodic source over a finite alphabet B such that $H(U) > H(X)$. Given any $\delta > 0$ there is a source $[B,\lambda,Y]$ and a stationary invertible code $\bar{f}: A^Z \to B^Z$ such that $Y = \bar{f}X$ and $d_w(\lambda,\tau) < \delta$.

Thus, X can be encoded arbitrarily close to any ergodic source with enough entropy (cfr. Smorodinsky (1971) or Shields (1979) for a discussion of the "enough entropy" condition). The proof is based on an extension to countable alphabets of the coding technique due to Kieffer (1981), Kieffer's construction in turn modifies that of Gray, Ornstein, and Dobrushin (1980).

In particular, let $K = \min\{k \in Z: \log_2 k > H(X)\} = INT(\exp_2 H(X)) + 1$. Take $B = \{1,2,\dots,K\}$ and then $[B,\tau,U]$ as the equiprobable IID source over B. Hence, $H(U) = \log_2 K > H(X)$, and Theorem 1 entails:

Corollary. Let $[A,\mu,X]$ be as in Theorem 1. For any $\delta > 0$ there is a source $[B,\lambda,Y]$ over an alphabet B with at most $INT(\exp_2 H(X))+1$ letters, and a stationary invertible code $\bar{f}: A^Z \to B^Z$ such that $Y = \bar{f}X$ and

$$\Sigma_{n=1}^{\infty} \Sigma_{y \in B^n} |\lambda^n(y) - K^{-n}| < \delta, \quad K = INT(\exp_2 H(X))+1 \;.$$

Thus, $\bar{f}$ accomplishes, in a stationary way, the aim of noiseless coding - the redundancy removal. At the same time, any stationary code $\bar{f}$

[1] Institute of Measurement and Measuring Technique, Electro-Physical Research Centre, Slovak Academy of Sciences, 885 27 Bratislava, ČSSR

induces a measurable map $f: A^Z \rightarrow B$ (viz. $fx = (\bar{f}x)_0$) and f in turn determines a partition of A^Z (viz. $P_f = \{f^{-1}\{b\} : b \in B\}$). Thus, our theorem generalizes (for shifts) Krieger's (1970) theorem on finite generators.

2. Distortion of Keane-Smorodinsky Codes

Let $[A, \mu, X]$ be an IID source over a finite alphabet A. Suppose $\hat{A}$ is a finite reproduction alphabet and ρ a finite-valued distortion measure on $(A \cup \hat{A}) \times (A \cup \hat{A})$. We assume that the image of $\hat{A}$ is contained in A, that is, for each $a \in A$ there is a unique $\hat{a} \in \hat{A}$ such that $\rho(a, \hat{a}) = 0$ (this condition excludes some unpleasant situations in rate distortion theory, see p.29 of Berger (1971), otherwise, it does not too much restrict the validity of source coding theorems as pointed out by Kieffer (1980)). Let $\bar{f}: A^Z \rightarrow \hat{A}^Z$ be a stationary code. Its rate is defined to be the entropy of the encoded process, i.e., $H(\bar{f}) = H(\bar{f}X)$, and its average distortion is, by definition,

$$\rho(\bar{f}) = E_\mu [\rho(X_0, (\bar{f}X)_0)].$$

For $0 < R < H(X)$ put

$$\delta(R, \mu) = \inf \{\rho(\bar{f}) : H(\bar{f}) \leq R\};$$

$\delta(., \mu)$ is said to be the optimum performance theoretically attainable (OPTA) using stationary codes. As pointed out by Kieffer (1980) and Gray, Neuhoff, and Ornstein (1975), $\delta(., \mu)$ coincides for ergodic sources with OPTA's using block and sliding-block codes (in the limit of increasing length). Moreover, for ergodic sources the coding theorem is valid saying that $\delta(R, \mu) = D(R)$, where $D(.)$ is the distortion rate function (cfr. Gray, Neuhoff, and Omura (1975)). Let now $[A, \mu, X]$ and $[\hat{A}, \nu, Y]$ be two stationary sources. Let

$$\bar{\rho}(X, Y) = \inf \{E_\pi [\rho(X_0, Y_0)] : \pi \in \mu \vee \nu\},$$

where $\mu \vee \nu$ is the set of all stationary pair sources over $A \times \hat{A}$ yielding μ and ν as marginals. This is a generalization of Ornstein's $\bar{d}$ distance. If both X and Y are IID, then one can consider only first-order marginals, that is, $\bar{\rho}(X, Y) = \bar{\rho}_1(X, Y)$; see Gray, Neuhoff, and Shields (1975). An alternate form of coding theorem involves the $\bar{\rho}$-distance, and asserts that

$$\delta(R, \mu) = \inf \{\bar{\rho}(X, Y) : H(Y) \leq R\}.$$

Our result is in spirit of this latter form. Let

$$\delta_e(R, \mu) = \inf \{\rho(\bar{f}) : H(\bar{f}) \leq R,\ \bar{f}X \text{ is IID}\};$$

$$\bar{\rho}(R, \mu) = \inf \{\bar{\rho}(X, Y) : H(Y) \leq R,\ Y \text{ is IID}\}.$$

Observe that if Y is any IID process with $H(Y) \leq R < H(X)$, a stationary code $\bar{f}: A^Z \rightarrow \hat{A}^Z$ can be constructed so that $Y = \bar{f}X$ and $\bar{f}$ has finite expected code length (cfr. Keane and Smorodinsky (1977)).

Theorem 2. Let $[A,\mu,X]$, $\hat{A}$, and ρ be as above. Then, for all $0<R<\mathbb{H}(X)$, we have that $\delta_e(R,\mu) = \bar{\rho}(R,\mu)$.

The proof is based on an approximation argument with the aid of the fact that Keane-Smorodinsky codes have finite expected code length. In fact, using this property we can find a length constraint so large that Keane-Smorodinsky sequential coding procedure stops by knowing a segment of a source sequence of length shorter than the constraint, for all but a set of sequences of arbitrarily small probability. The restriction of a minimal global assignment to skeletons of that constrained length always can be modified so that it satisfies also a properly formulated average distortion constraint.

We conjecture that $\delta_e(R,\mu) = \delta(R,\mu)$, i.e., the OPTA defined in terms of IID encoded processes is the same as the OPTA using codes which result in B-processes. The latter clause is justified by the fact that any stationary code is a homomorphism of the corresponding shifts, and any factor of an IID process is a B-process; see Ornstein (1974), for example. If true, the conjecture would have interesting consequences from the point of view of decryption complexity.

3. Block Codes and The Isomorphism Problem

Let $[A,\mu,X]$ be a stationary non-ergodic source over a finite alphabet A with ergodic decomposition

$$\mu(.) = \int_{R_A} \mu_x(.)\mu(dx).$$

Let $\Delta_n \in E_n(\varepsilon)$ if $\Delta_n \subset A^n$ and $\mu^n(\Delta_n) > 1-\varepsilon$. We call the limit (if it exists)

$$H_\varepsilon(\mu) = \lim_{n\to\infty} n^{-1}\log_2 \min\{\mathrm{card}(\Delta_n): \Delta_n \in E_n(\varepsilon)\}$$

the epsilon-rate of $[A,\mu,X]$ (cfr. Šujan (1977,1980)).

Theorem 3. Let $[A,\mu,X]$ and $[B,\nu,Y]$ be two exchangeable finite alphabet sources (that is, each of the sources has only IID components in its ergodic decomposition). If $H_\varepsilon(\mu) = H_\varepsilon(\nu)$ for all $\varepsilon \in (0,1)$ for which the limits exist, then there is a stationary invertible code $\mathbf{F}: A^Z \to B^Z$ such that $Y = \mathbf{F}X$.

The idea of the proof is as follows. Our assumption entails that the ergodic components of the two sources can be divided into pairs (one member of that pair for one source) of IID sources of the same entropy, hence, these components are isomorphic by Ornstein's (1970) result. The rest is based on a refinement of a construction due to Winkelbauer (1977) to the effect that the "local" isomorphisms give rise to a "global" one between the mixtures. The details of the proof as well as of the other proofs will appear elsewhere.

References

Berger,T.(1971). Rate Distortion Theory. Prentice Hall, Englewood Clif.

Gray,R.M.,Neuhoff,D.L.,Omura,J.K.(1975).Process definitions of the distortion rate function and source coding theorems. IEEE Trans.Inform. Theory IT-21,524-532.

Gray,R.M.,Neuhoff,D.L.,Ornstein,D.S.(1975).Nonblock source coding with a fidelity criterion. Ann.Prob.3, 478-491.

Gray,R.M.,Neuhoff,D.L.,Shields,P.C.(1975).A generalization of Ornstein's $\bar{d}$ distance with applications to information theory. Ann. Prob. 3, 315-328.

Gray,R.M.,Ornstein,D.S.,Dobrushin,R.L.(1980). Block synchronization, sliding-block codes, invulnerable sources, and zero error codes for discrete noisy channels. Ann.Prob. 8, 639-674.

Keane,M.,Smorodinsky,M,(1977). A class of finitary codes. Israel J. Math. 26, 352-371.

Kieffer,J.C.(1980). Extension of source coding theorems for block codes to sliding-block codes. IEEE Trans.Inform.Theory IT-26,679-692.

Kieffer,J.C.(1981). Zero error stationary coding over stationary channels.Z.Wahrscheinlichkeitstheorie verw.Geb. 56, 113-126.

Krieger,W.(1970). On entropy and generators of measure preserving transformations. Trans. Amer.Math.Soc. 149, 453-464.

Ornstein,D.S.(1970).Bernoulli shifts with the same entropy are isomorphic. Adv. Math. 4, 337-352.

Ornstein,D.S.(1974).Ergodic Theory, Randomness, and Dynamical Systems. Yale University Press, New Haven and London.

Shields,P.C.(1979).Stationary coding of processes. IEEE Trans. Inform. Theory IT-25, 283-291.

Smorodinsky,M.(1971). Ergodic Theory, Entropy. Lect.Notes Math. 214. Springer, New York.

Šujan,Š.(1977). A generalized coding problem for discrete information sources. Supplement, 95 pp. Kybernetika 13.

Šujan,Š.(1980). Epsilon-rates, epsilon-quantiles, and group coding theorems for finitely additive information sources. Kybernetika 16, 105-119.

Winkelbauer,K.(1977). On the existence of finite generators for invertible measure-preserving transformations. Comment. Math. Univ. Carolinae 18, 789-812.

ON STOCHASTIC MAPS OF C^*-ALGEBRAIC STATE SPACES

Armin Uhlmann[1)]

1. Some notations, definition of stochastic maps.

We consider C^*-algebras, A, B, ... , with an identity which we denote by $\mathbb{1}$ or, more carefully, by $\mathbb{1}_A$, $\mathbb{1}_B$, ... Let A^* denote the Banach space of bounded linear functionals of the algebra A . We need the following subsets of A^* : The set of Hermitian functionals, A^*_h , the cone of positive linear functionals, A^*_+ , and the convex set of states, Ω_A . In particular we have $u \in \Omega_A$ iff $u \in A^*_+$ and $u(\mathbb{1}_A) = 1$.

Given two C^*-algebras, A and B , the Banach space of the bounded linear maps

$$T : A^* \longrightarrow B^*$$

is denoted by $\mathrm{Lin}(A^*,B^*)$. In this space there is a natural weak topology, the neighbourhoods of which around the identity map are given by

$$\left\{ T : |(Tu_1)(b_1)| + \dots + |(Tu_m)(b_m)| < \varepsilon \right\}$$

where $u_1,\dots,u_m$ are from A^* , and $b_1,\dots,b_m$ are elements of B . One knows that a weakly closed and bounded in the Banach norm of Lin(..) set is a weakly compact one.

A map $T: A^* \longrightarrow B^*$ is called positive (or positive preserving) iff

$$T : A^*_+ \longrightarrow B^*_+$$

The linear map $T : A^* \longrightarrow B^*$ is called <u>stochastic</u> iff it maps Ω_A into Ω_B ,

$$T : \Omega_A \longrightarrow \Omega_B$$

Stochasticity of T is equivalent with linearity, positivity preserving, and the normalizing condition $(Tu)(\mathbb{1}_B) = u(\mathbb{1}_A)$. We call

$$\mathrm{St}(A^* , B^*)$$

the set of all stochastic maps from A^* into B^* .
This set is a weakly compact convex subset of $\mathrm{Lin}(A^*,B^*)$.

We shall distinguish a subclass of stochastic maps. By

$$\mathrm{St}^e(A^* , B^*)$$

we denote the weak closure of the set of all maps T allowing for a representation

$$\forall\, u \in A^* : \quad T u = \sum_j u(a_j)\, v_j \quad \text{, finite sum ,}$$

where $v_1, v_2, \dots$ are states of the algebra B , i.e. out of Ω_B, and $a_1, a_2, \dots \in A$ such that $a_1 + a_2 + \dots = \mathbb{1}_A$, and $a_j \geq 0$ for all j .

1) Dep. of Physics and NTZ, Karl-Marx-University, Leipzig, GDR .

2. An inequality.

At first we consider real-valued functions, $f = f(s_1,\dots,s_n)$, defi= ned on R^n and satisfying

$$\text{(a)} \quad \forall \quad s \geqslant 0 : \quad f(ss_1,\dots,ss_n) \;=\; s\, f(s_1,\dots,s_n)$$

$$\text{(b)} \quad f(s_1+t_1,\dots,s_n+t_n) \;\leqslant\; f(s_1,\dots,s_n) \;+\; f(t_1,\dots,t_n)$$

For the purpose of abbreviation we refer to such functions as to h-convex functions. (Under the condition (a) the subadditivity (b) and convexity are equivalent.)

Let now A be a C^*-algebra with identity element. Choosing on R^n an h-convex function f we define for every n-tuple of Hermitian functionals $u_1,\dots,u_n \in A_h^*$ the expression

$$S_f(u_1,\dots,u_n) \;:=\; \sup \sum_j f(\, u_1(a_j),\; \dots,\; u_n(a_j)\,) \tag{1}$$

where the supremum is running through all the partitions of the identity $\{a_1,\dots,a_m\}$ of A , i.e. through all possible choices of finitely many elemente $a_j \in A$ with $a_j \geqslant 0$ and $a_1+a_2+\dots+a_m = 1_A$,

The subadditivity of f is guarantying the finiteness of S_f .

Theorem 1: Let $u_1,\dots,u_n$ denote n Hermitian functionals from A_h .
Then we have

$$S_f(\,Tu_1,\dots,\,Tu_n) \;\leqslant\; S_f(u_1,\dots,u_n) \tag{2}$$

for every $T \in St(A^*,B^*)$, and for every h-convex f .

To prove this we need

Lemma 1: Let $\tilde{u}_1,\dots,\tilde{u}_n$ be the uniquely defined normal extensions of $u_1,\dots,u_n$ to the bidual A^{**} of A . Then

$$S_f(\tilde{u}_1,\dots,\tilde{u}_n) \;=\; S_f(u_1,\dots,u_n) \tag{3}$$

This statement (see p.53 of [1]) is a simple consequence of the fact that every finite partition of the identity of A^{**} can be approximated by those of A in the strong operator topology of A^{**} , see lemma 2.1.2. of [1] where this lemma is reduced to the KAPLANSKI density theorem (compare [2]). Now using the lemma above we proceed as follows. $T\colon A^* \longrightarrow B^*$ induces $T^*\colon\ B^{**} \to A^{**}$ and $T^{**}\colon\ (A^{**})^* \longrightarrow (B^{**})^*$. On the imbeddings

$$A^* \hookrightarrow (A^{**})^* \qquad \text{and} \qquad B^* \hookrightarrow (B^{**})^*$$

given by the normal extensions, T^{**} coincides with T and because of lemma 1 it suffices to show the inequality

$$S_f(\,T^{**}\tilde{u}_1,\dots,T^{**}\tilde{u}_n\,) \;\leqslant\; S_f(\tilde{u}_1,\dots,\tilde{u}_n)$$

Let us choose $\delta > 0$ and a partition of the identity $\{b_1,\dots,b_m\}$ of B^{**} such that

$$\delta \;+\; S_f(T^{**}\tilde{u}_1,\dots,T^{**}\tilde{u}_n) \;\leqslant\; \sum f(\,(T^{**}\tilde{u}_1)(b_j),\dots,(T^{**}\tilde{u}_n)(b_j)\,)$$

The right hand side equals

$$\sum_j f(\tilde{u}_1(T^* b_j),\ldots,\tilde{u}_n(T^* b_j)) \qquad (x)$$

However, T^* is a positivity preserving map from B^{**} into A^{**} and it is $T\,\underline{1}_{B^{**}} = \underline{1}_{A^{**}}$. Therefore, the expression (x) is smaller than $S_f(\tilde{u}_1,\ldots,\tilde{u}_n)$. /=/

We may use the proof to obtain a further statement. Let T be positive and contracting from A^* into B^*. Then T^* is positive and contracting from B^{**} into A^{**}. Hence $\underline{1} - \sum_j T^* b_j = a$ is a positive element. Hence after adding $f(\tilde{u}_1(a),\ldots,\tilde{u}_n(a))$ to (x) in this new situation the result will be smaller than $S_f(\tilde{u}_1,\ldots,\tilde{u}_n)$. The additional term is non-negative provided $\forall\, j$: $u_j \in A^*_+$, and $f \geqslant 0$ in case of $\forall\, j$: $s_j \geqslant 0$. Denoting by R^n_+ the subset of R^n given by $\forall\, j$: $s_j \geqslant 0$ one gets

<u>Theorem 2:</u> Let $T : A^* \longrightarrow B^*$ be a positive contraction. Let f be h-convex and assume $f \geqslant 0$ on $R_+{}^n$.
Then for every n-tuple of positive linear functional $u_1,\ldots,u_n \in A^*_+$ it is

$$S_f(Tu_1,\ldots,Tu_n) \leqslant S_f(u_1,\ldots,u_n)$$

<u>Remark:</u> Let f be h-convex. Then $f \geqslant 0$ on $R_+{}^n$ is equivalent with

$$f(s_1',\ldots,s_n') \geqslant f(s_1,\ldots,s_n) \quad \text{if } \forall\, j : s_j' \geqslant s_j .$$

Let us hence call

<u>h^+-convex</u> resp. <u>h^--convex</u>

every h-convex function f on R^n satisfying

$f \geqslant 0$ on $R_+{}^n$ resp. $f \leqslant 0$ on $R_+{}^n$

3. Existence of stochastic maps.

We start by constructing certain functionals of type S_f.
Let B be a C^*-algebra and $b_1,\ldots,b_n$ Hermitian elements of it. Given real numbers $s_1,\ldots,s_n$, the norm of the positive part of $s_1b_1+\ldots+s_nb_n$,

$$f := \| (s_1b_1+\ldots+s_nb_n)_+ \| = \sup_{v \in \Omega_B} v(s_1b_1+\ldots+s_nb_n) \qquad (4)$$

defines, considered as a function on R^n , an h-convex function.
The more, in case $\forall\, j$: $b_j \geqslant 0$, it follows h^+-convexity for f , and in case $\forall\, j$: $b_j \leqslant 0$ it will follow h^--convexity of f .
Let us now consider a further C^*-algebra, A , and $u_1,\ldots,u_n$ some of its Hermitian linear functionals. We write

$$K(u_1,\ldots,u_n;b_1,\ldots,b_n) := S_f(u_1,\ldots,u_n) \quad \text{with } f = \| (s_1b_1 + \ldots + s_nb_n)_+ \| \qquad (5)$$

These particular S_f-functionals we need in the following.

We rewrite the question, whether there is an inverse to theorem 1 statement in terms of the algebras

$$A^{(n)} := A \oplus A \oplus \dots \oplus A \quad \text{and} \quad B^{(n)} := B \oplus \dots \oplus B\ ,$$

with n direct summands. Of course, $A^{(n)}$ is a C^*-algebra again, and

$$A^{(n)*} = A^* \oplus \dots \oplus A^*\ , \quad n \text{ direct summands}\ ,$$

in a natural way. For technical reasons only let us denote by

$$I_+ \quad \text{and} \quad I_-$$

the following sets of maps

$$T^{(n)} : \quad A^{(n)*} \longrightarrow B^{(n)*}$$

which are constructed this way: There is $T \in St^e(A^*,B^*)$, and there are $w_1,\dots,w_n \in B_h^*$ such that for all $u_1,\dots,u_n \in A^*$ we have

$$T^{(n)} : \ u_1 \oplus \dots \oplus u_n \longrightarrow (Tu_1+w_1) \oplus \dots \oplus (Tu_n+w_n)\ .$$

Using this setting we demand

$$T^{(n)} \in I_+ \quad \text{iff} \quad \forall j : \quad w_j \geq 0$$

$$T^{(n)} \in I_- \quad \text{iff} \quad \forall j : \quad -w_j \geq 0\ .$$

The concept of weak topology remains relevant in an obvious manner for transformations from $A^{(n)*}$ into $B^{(n)*}$ not being linear necessarily. I_+ and I_- are composed of a weakly compact and a weakly closed set of transformations, and they are both weakly closed, hence. Therefore the sets

$$F_+ := I_+ (u_1 \oplus \dots \oplus u_n) \quad \text{and} \quad F_- := I_- (u_1 \oplus \dots \oplus u_n)$$

of $B^{(n)*}$ are weakly closed convex ones.

Let us at first consider F_+ , and let us assume $x_1 \oplus \dots \oplus x_n \notin F_+$. Then there exists a weakly continuous Hermitian linear form q on $B^{(n)*}$ separating this element from F_+ , i.e.

$$q(\, x_1 \oplus \dots \oplus x_n\,) > \sup_{r \in I_+} q(r)\ .$$

Being Hermitian and weakly closed there are Hermitian elements $b_1,\dots,b_n$ out of B such that

$$\forall\ v_1 \oplus \dots \oplus v_n \quad B_h^{(n)*} : \ q(v_1 \oplus \dots \oplus v_n) = v_1(b_1)+\dots+v_n(b_n).$$

With this choice of $b_1,\dots,b_n$ we get

$$x_1(b_1)+\dots+x_n(b_n) > \sup (Tu_1)(b_1)+\dots+(Tu_n)(b_n) + w_1(b_1)+\dots+w_n(b_n).$$

By definition, $w_1,\dots,w_n$ are varying freely within B_+^* . Therefore the supremum can and will be finite for $b_1 \leq 0,\ \dots,\ b_n \leq 0$, only.

But then the supremum will be reached with $w_1 = w_2 = \dots = 0$ only.

Thus we get

$$x_1(b_1)+\dots+x_n(b_n) > \sup_{T \in St^e(A^*,B^*)} (Tu_j)(b_j)\ .$$

But $St^e(A^*,B^*)$ is the weak closure of certain known in their structure maps, and in performing the desired supremum it suffices to take only these. Hence the supremum has to run through all decompositions, $a_1,\dots,a_m$, of $\underline{1}_A$ with $m = 1,2,\dots$, and all choices of m states of B in

$$x_1(b_1)+\dots+x_n(b_n) > \sup \sum_i \sum_j u_j(a_i)\, v_i(b_j) .$$

Performing the supremum with respect of the states $v_1,\dots,v_m \in \Omega_B$ we get

$$x_1(b_1)+\dots+x_n(b_n) > \sup \sum_i \| (\sum_j u_j(a_i)\, b_j)_+ \| .$$

The right hand side of this inequality equals $K(u_1,\dots,u_n;b_1,\dots,b_n)$. We see: $x_1 \oplus \dots \oplus x_n \notin F_+$ if there are $b_1 \leq 0, \dots, b_n \leq 0$, fulfilling this inequality. But the reverse, $y_1 \oplus \dots \oplus y_n \in F_+$, may be expressed by

$$\forall j : \; y_j = T u_j + w_j , \quad w_j \geq 0 , \quad T \in St^e(A^*,B^*) .$$

Thus we have established (we pass from $b_j \leq 0$ to $-b_j$, $b_j \geq 0$):

<u>Theorem 3:</u> Let A, B be two C^*-algebras with identity elements, and $u_1,\dots,u_n \in A_h^*$ and $v_1,\dots,v_n \in B_h^*$.
There is a map $T \in St^e(A^*,B^*)$ satisfying

$$\forall j = 1,2,\dots,n: \quad T u_j \leq v_j$$

if and only if for all elements $b_1 \geq 0,\dots, b_n \geq 0$ out of B

$$- [v_1(b_1) + \dots + v_n(b_n)] \leq K(u_1,\dots,u_n;-b_1,\dots,-b_n)$$

is valid.

Literally the same reasoning with F_- instead of F_+ gives

<u>Theorem 4:</u> Let $u_1,\dots,u_n \in A_h^*$ and $v_1,\dots,v_n \in B_h^*$ be given where A and B are C^*-algebras with identity elements.
There is a map $T \in St^e(A^*,B^*)$ satisfying

$$\forall j : \quad T u_j \geq v_j$$

iff for all $b_1 \geq 0,\dots, b_n \geq 0$ out of B we have

$$v_1(b_1) + \dots + v_n(b_n) \leq K(u_1,\dots,u_n; b_1,\dots,b_n) .$$

<u>Corollary:</u> Assuming

$$\forall j : \quad u_j \in \Omega_A , \quad v_j \in \Omega_B ,$$

the existence of a stochastic map out of $St^e(A^*, B^*)$ with $\forall j$: $T u_j = v_j$ is equivalent with

$$v_1(b_1)+\dots+v_n(b_n) \leq K(u_1,\dots,u_n;b_1,\dots,b_n)$$

where $b_1,\dots,b_n \in B$ runs either through all positive elements, or through all negative elements, or through all Hermitian elements of B.

This, for commutative A and B, has been proved in the same manner by Alberti and Uhlmann [1] as an intermediate result in order to prove the inverse to theorem 1 statement. It turns out that only the commutativity of B is needed for this purpose, and, indeed, an even slightly weaker assumption. It is our next aim to show this.

Definition: Let B be a C^*-algebra with identity element. We shall say the n-tuple $\{v_1,\dots,v_n\}$, $\forall j : v_j \in B_h^*$, fulfils condition (C), iff for every choice of n Hermitian elements, $b_1,\dots,b_n$, out of B one has

$$v_1(b_1) + \dots + v_n(b_n) \leq K(v_1,\dots,v_n;b_1,\dots,b_n) . \tag{6}$$

We now see the following: The validity of equ. (2) and (6) implies because of (5) the validity of the assumptions used in theorems 3 and 4. Together with theorem 1 the following is nothing but a mere re-writing of the results already obtained:

Theorem 5: Let A and B be two C^*-algebras with identity element. Let $u_1,\dots,u_n \in A_h^*$ and $v_1,\dots,v_n \in B_h^*$ be two n-tuples of Hermitian linear forms, and assume the validity of condition (C) for the n-tuple $\{v_1,\dots,v_n\}$.
Then there exists a stochastic map $T \in St^e(A^*,B^*)$ satisfying

(a) $\forall j : T u_j \succ v_j$

if and only if for all h^+-convex functions, f, it is

$$S_f(u_1,\dots,u_n) \succ S_f(v_1,\dots,v_n)$$

(b) $\forall j : T u_j \leq v_j$

if and only if for all h^--convex functions, f, it is

$$S_f(u_1,\dots,u_n) \geq S_f(v_1,\dots,v_n) .$$

We shall now supplement these statements by a more handly than condition (C) assumption.

Lemma 2: Let B be a commutative C^*-algebra with identity. Then every n-tuple of Hermitian linear forms $v_1,\dots,v_n \in B_h^*$ fulfils the condition (C).

This lemma is an important particular case of the following

Conjecture: Let B be a C^*-algebra with identity element, and $\{v_1,\dots,v_n\}$ a n-tuple out of B_h^*.
Let B_o be the following C^*-subalgebra of B : $b \in B$ is an element of B_o if and only if for every $c \in B$ we have

$$v_j(c\, b) = v_j(b\, c) , \quad j=1,2,\dots,n . \tag{7}$$

Then $\{v_1,\dots,v_n\}$ satisfies condition (C) if B_o contains a maximal commutative subalgebra of B.

From theorems 3 and 4 one concludes the validity of condition (C) if and only if

$$\forall j = 1,\dots,n : \quad T v_j = v_j \quad \text{with suitable} \quad T \in St^e(B^*,B^*) \qquad (8)$$

can be fulfilled. For commutative B this has been proved (see theorem 2.3.3 of [1]) by showing

$$\text{if } B \text{ is commutative then: } St(B^*,B^*) = St^e(B^*,B^*) . \qquad (9)$$

In this particular case the identity map is contained in St^e by (9), and we are allowed to take in (8) the identity map.

<u>Lemma 3:</u> Let B be a W^*-algebra, $D \subseteq B_o$ a discrete maximal abelian subalgebra, and $v_1,\dots,v_n$ normal functionals satisfying (7). Then the conjecture is true, i.e. $\{v_1,\dots,v_n\}$ fulfils condition (C).

'D is discrete' is understood as following: There is a family $\{p_k\}$, indexed by an index set, of projections satisfying

(i) $\sum p_k$ = identity map, where the convergence is the weak one over the directed set of finite subsets of the index set.

(ii) $\forall b \in B, \forall k : \quad p_k b p_k = \beta_k(b) p_k$ with complex numbers $\beta_k(p)$.

Let now N be a finite subset of the index set. Defining the projection q_N by $\underline{1} = q_N + \sum p_k$, summation over N, we get a resolution of the identity. Let us choose a state u such that $u(q_N) \neq 0$, and $\forall k \in N : \; u(p_k) \neq 0$. Then we define T_N by

$$(T_N v)(b) = u(q_N)^{-1}u(q_N b q_N)v(q_N) + \sum u(p_k)^{-1}u(p_k b p_k)v(p_k) \qquad (10)$$

where the summing is over N. Obviously, $T_N \in St^e(B^*,B^*)$.
Let v be normal. Given $\varepsilon > 0$ there is a finite subset N of the index set such that the norms of $v - \sum_N v(p_k \cdot)$ and $v(q_N \cdot q_N)$ are smaller than ε. Because of (ii) it is

$$u(p_k)^{-1}u(p_k b p_k)v(p_k) = \beta_k(b)v(p_k) = v(p_k b p_k) .$$

If, therefore, $v(bc) = v(cb)$ for all $c \in D$, the sum of the right-hand side of (10) equals $v(b) - v(bq_N)$. Hence $|(T_N v)(b) - v(b)| < 2\varepsilon\|b\|$. From this we infer the existence of a weak limit T of the operators T_N with $T v = v$ for all normal states satisfying the assumption of the conjecture. Thus lemma 3 is proved. /=/

Let us now assume B to be a C^*-algebra with unit element, and D a maximal commutative *-subalgebra. Let us define

$$Q = \{ v \in B_h : v(bd) = v(db) \text{ for all } b \in B, d \in D \} . \qquad (11)$$

We denote by v_D the restriction of a linear functional $v \in B^*$ onto D, and $\|v_D\|$ is its functional norm, i.e. its norm as an element of the Banach space D^*.

Theorem 6: Under the condition that

$$\forall v \in Q: \quad \|v\| = \|v_D\| \tag{12}$$

every finite subset of Q satisfies condition (C).

Proof: Step 1: Let $w \in D_h^*$. There is a $v \in Q$ with $v_D = w$. We need to prove this only for states. Let w be a state of D and $R(w)$ the set of all those states v of B the restriction on D of which is w, i.e. $v_D = w$. $R(w)$ is not empty by well known theorems [2]. Further, $R(w)$ is a weakly compact convex set. The map $v(.) \longmapsto v(e^{-1}.e)$, where e denotes a unitary element of D, is a continuous automorphism of $R(w)$. If e runs through the unitaries of D these maps are commuting one with another. Therefore, by the KAKUTANI fixed point theorem, there is $v_o \in R(w)$ with $v_o = v_o(e^{-1}.e)$ for all unitaries $e \in D$. From this it follows $v_o \in Q$.

Step 2: If $v, \tilde{v} \in Q$ and $v_D = \tilde{v}_D$, then $v = \tilde{v}$. Proof: condition (12) implies

$$\|v - \tilde{v}\| = \|v_D - \tilde{v}_D\| = 0 \quad \text{for} \quad v_D = \tilde{v}_D .$$

Step 3: Let

$$T_\beta : \quad (T_\beta w)(d) = \sum w_j(d)\, w(d_j)$$

with states w_j of D and $d_1, d_2, \ldots$ a decomposition of the unity in D. Then we have $T_\beta \in St^e(D^*, D^*)$.

As already mentioned there is a net $\{T_\beta\}$ with $\lim T_\beta =$ id. of D_h^* in the weak topology. Let v_j be the unique (!) extension of w_j to B which is in Q, $v_j \in Q$, $(v_j)_D = w_j$. Then

$$\hat{T}_\beta = \sum v_j(\,.\,)\, w(d_j)$$

extends T_β to $St^e(B^*, B^*)$. Let $\hat{T}$ be a weak limit of the net $\{\hat{T}_\beta\}$. Then for all $v \in Q$ we have $(\hat{T}v)_D = v_D$ and $\hat{T}v \in Q$. But the unique extension of v_D is v. This implies $\hat{T}v = v$ for all $v \in Q$. This, together with $\hat{T} \in St^e(B^*, B^*)$, implies the validity of condition (C) for all finite subsets of Q. /=/

Remark: If B is commutative then $St(A^*, B^*) = St^e(A^*, B^*)$, and every such map is a completely positive one (in the sense of Umegaki and Stinespring). The same, i.e. the complete positivity, is true for the map constructed in proving lemma 3.

Remark: For pairs of Hermitian functionals ($n = 2$) and with $A = B$ the n-dimensional commutative *-algebra of all complex-valued functions on n points our theorems are equivalent with various classical results concerning the existence of stochastic and doubly stochastic matrices. They are well described in [3 - 8]. A more recent contribution is in [9] and, besides [1] and the references cited there, in [10].

Acknowledgement

It is a pleasure to thank Peter M. Alberti for many stimulating and critical discussions. The line of arguments used in this paper would not be a complete one without his valuable contributions.
In addition I like to thank several participants of the "Mathematisch-Physikalisches Seminar der Karl-Marx-Universität".

References

[1] Alberti,P.M.; Uhlmann,A.: Dissipative Motion in State Spaces. Teubner-Texte zur Mathematik Bd.33, Leipzig 1981.

[2] Sakai,S.: C^*-algebras and W^*-algebras. Springer-Verl., Berlin, Heidelberg, N.Y., 1971

[3] Hardy,G.H.; Littlewood,J.E.;Polya,J.E.; : Inequalities. Cambridge Univ. Press, N.Y., 1934.

[4] Horn,A.: Proc.Nat.Acad.Sci. USA 36 (1950) 374

[5] Karamata,H.: Publ.Math.Univ.Belgrade, 1 (1932) 145

[6] Ostrowski,A.: J. Math.Pures et Appl. 31 (1952) 253

[7] Rado,R.: J. London Math.Soc. 27 (1952) 1

[8] Sherman,S.: Proc.Nat.Acad.Sci. USA 37 (1951) 826

[9] Ruch,E.; Schranner,R.;Seligman,Th.: J. Chem.Phys. 69 (1978) 386

[10] Alberti,P.M.; Uhlmann,A.: Stochasticity and Partial Order. VEB Deutsch.Verl.der Wissensch., Berlin, 1981.

A THEOREM ON PERIODICAL MARKOV APPROXIMATION IN ERGODIC THEORY

A. M. VERŠIK[1]

1. Introduction

In the paper [1] a proof is sketched of the theorem on simultaneous uniform approximation of the operators of multiplication and shift on the space of L^2-functions defined on a dynamical system with invariant measure, more exactly: let S be an automorphism of the interval [0,1] with invariant Lebesgue measure and φ a bounded measurable function, then for any $\varepsilon > 0$ there exist operators $U^{(\varepsilon)}$ and $V^{(\varepsilon)}$ on L^2 such that $\|U_S - U^{(\varepsilon)}\| < \varepsilon$, $\|M\varphi - V^{(\varepsilon)}\| < \varepsilon$ and the $*$-algebra, generated by $U^{(\varepsilon)}$ and $V^{(\varepsilon)}$, is finitedimensional, U_S denotes the unitary operator of S, $M\varphi$ denotes the multiplication operator and $\|.\|$ the operator norm. If $(Sf)(x) = f((x+\alpha) \bmod 1)$, $\varphi = \exp i2\pi x$, then one deals with the simultaneous approximation of operators satisfying the relation $UMU^{-1}M^{-1} = \exp i2\pi\alpha I$. This problem was solved in [2] and posed earlier by the author in [3,4] and by Rieffel in [5]. The above formulated theorem has applications in the theory of operators, in the approximation theory of dynamical systems and in the theory of C^*-algebras; it may be considered as a variant of noncommutative constructive operator theory (see [4]). The proof of the theorem is divided into two parts. The first represents a new variant of the theory of periodical approximation (theorem 3 in [1]) by constructing a Markov tower of periodical automorphisms (see below). This part will be considered in this article. The second part belongs to the theory of AF-algebras and will be elaborated elsewhere. In the theory of dynamical systems and in the theory of equations approximation problems may be divided into the two classes of factor approximation or momentum methods (projective approximation) and of subapproximation or net methods (inductive approximation). In the first case we

1) Leningrad State University, Mathematics Institute,
19 10 28 Leningrad

consider the behaviour of operators on finitedimensional subspaces of functions or on set systems, in the second case the behaviour of the system on finite point sets. The further considered approximation is of the second kind, we approximate the automorphism by periodical automorphisms whose trajectories are contained in the trajectories of the original one, i.e. we approximate the automorphism by a random permutation. The idea of the considered variant touches the author's paper [6] on the scale of an automorphism and uses a modification of the construction there applied ([6], § 3, Lemma) of a corresponding sequence of periodical automorphisms. An important modification introduced here consists in the fact that we do not desire the unalterability of the minimal periods over all points for all periodical automorphisms of the approximation. Instead of this the correspondence of the approximation steps is stronger: any permutation in only one point of its layer differs from the approximated automorphism and as a function on the quotient space with respect to the trajectories of the periodical automorphism with values in the (alternating) group of permutations is measurable with respect to a certain partition, which is complementary to the trajectorial partition of the following approximation term (see below). Such a hard demand excludes in many ways an additional arbitrariness of approximation as usually possible. We also get to the notion of Markov tower or of adic model of an automorphism, whose existence is also shown in this paper. Its construction or approximation in concrete cases is not easy; in § 4 we deal with some examples. In 2 the definitions and in § 3 the proof of the basic theorem are given. We recall the fact that the existence of a periodical automorphism that approximates a given automorphism uniformly (in the sense of measure theory) is the content of the classical Rochlin-Halmos lemma; the problem of consecutive corresponding periodical approximations was considered in different manners in a series of papers (see e.g. the survey [7]), projective approximations (multivalued mappings) are introduced in [8]. We also remark that in our point of view such approximation (multivalued, see [9])(in their metrical variant) should remain a subject for further study also in hydrodynamics.

2. Markov compact and adic transformations

In the following we construct a tower of periodical automorphisms for every automorphism of a Lebesgue space. But at first we define a corresponding model of the space and of its transformation, to which we want to transfer (i.e. construct an isomorphism) such a pair, i.e. we define a class of canonical towers.

Let $r_0=1$, $r_1,\dots$ be a sequence of natural numbers, $r_i \geq 2$, $i=1,\dots,$ M_1, $M_2,\dots$ a sequence of $\mathbb{Z}_+$ - matrices, where the order of the matrix $M_i = (m^i_{jk})$ is $r_{i-1}\times r_i$ ($i=1,2,\dots$). We suppose that all matrices have no zero-rows and -columns. We consider a graduated graph $\Gamma(\{M_i\})$, the vertices of which $D = \bigcup_{i=0}^{\infty} D_i$ are divided into finite 'stories' $D_0 = \{\emptyset\}$, $D_1,\dots$, where the number of elements of D_i is $|D_i| = r_i$ ($i = 0,1,2,\dots$). The edges of the graph join the vertices of the neighbourstories only and the number of the edges joining the vertex j of the (i-1)-th story and the vertex k of the i-th story equals m^i_{jk} ($k = 1,\dots,r_i$; $j = 1,\dots,r_{i+1}$; $i = 0,1,\dots$). From the conditions the matrices fulfil follows, that every vertex is joined at least with one vertex of the following story and (except $\emptyset \in D_0$) with one of the story before. We call a sequence of edges $(\gamma_0, \gamma_1,\dots)$, where γ_{i-1} and γ_i have a common vertex in the i-th story ($i=1,2,\dots$), a way in the graph. Let $\mathcal{X}(\Gamma)$ be the space of all ways of the graph $\Gamma(\{M_i\}_{i=1}^{\infty})$ with the weak topology. (A neighbourhood of a way is the set of all ways equal to the given way up to one place. It is evident that $\mathcal{X}(\Gamma)$ is a compact).

<u>DEFINITION 1</u>: $\mathcal{X}(\Gamma)$ with $\Gamma = \Gamma(\{M_i\})$ is called a Markov compact with parameters $\{r_i, M_i \ (i = 1,2,\dots)\}$. If $r_i \equiv r$ and $M_i \equiv M$ ($i = 1,2,\dots$), then the Markov compact is called stationary.

In topological dynamics often stationary Markov compacts and its two-side shifts are considered. We shall study arbitrary Markov compacts and its transformations of another form. The various properties of the compact $\mathcal{X}(\Gamma)$ and of the measures on this compact in dependence on the property of the matrices M_i to be asymptotical or not will be studied at another place. They are linked especially with the properties of AF-algebras, constructed according to the graph Γ how to the Bratelli-diagram [10].

We consider only the case that the elements of M_i are zeros and ones. In this case the graph has no multiple edges and every way is a sequence of vertices. This assumption does not restrict the generality, since by introduction of new vertices it is possible to get rid from multiplicity. In this case the Markov compact can be described in the usual way: it is a closed subset of the compact $\prod_{i=1}^{\infty} D_i$, consisting of all sequences $\{d_i\}$ for which $m^i_{d_i d_{i+1}} = 1$.

We give a direct description of the structure of a Markov compact.

STATEMENT 1: Every Markov compact is a totally disconnected separable compact X with an excluded sequence of finite partitions $\{\eta_i \ (i = 1,2,\dots)\}$, which has the properties

1) $\bigvee_{i=1}^{\infty} \eta_i = \varepsilon$

2) for every $k \geq 1$ and arbitrary set D, where D is an element of the partition η_k the correspondence $D \ni x \longrightarrow \{\eta_i(x);\ i \neq k\}$ defines a homeomorphism $D \simeq D/\bigvee_{i=1}^{k-1} \eta_i \times D/\bigvee_{i=k+1}^{\infty} \eta_i$.

Here $\bigvee$ is the sign of the refinement of partitions, i.e. of the partition consisting of all intersections of the elements of the refining partitions; ε is the partition into separate points. Condition 1 means, that for $x \neq y$ η_i exists such that $\eta_i(x) \neq \eta_i(y)$, where $\eta(x)$ is the element of η which contains x. In condition 2 the sign D/ξ denotes the factorisation of D by the partition ξ. The sense of condition 2 lies in the following: if $x,y \in D$, then $z \in D$ exists such that its projections on $D/_{\eta_i}$ for $i<k$ are the same as the projections of x, but on $D/_{\eta_i}$ for $i>k$ the same as the projections of y. If the number of elements of η_i is denoted by r_i and the matrix of intersections of η_{i-1} and η_i by M_i, then the compact X from this statement passes over to $\mathfrak{X}(\Gamma(\{M_i\}))$. Inversely, denoting by η_i the partition of $\mathfrak{X}$ into classes of ways, going through given vertex of the i-th story, we make sure of the fulfilment of the conditions of the statement. We will use the definition of the statement.

A subset of the Markov compact $\mathfrak{X}$, measurable with respect to $\bigvee_{i=1}^{k} \eta_i$ for any k, is called cylindrical. These subsets form an algebra, denoted by $\mathfrak{F}$. A transformation which lets $\bigvee_{i=1}^{k} \eta_i$ invariant and fixes $\bigvee_{i=k+1}^{\infty} \eta_i$ is called k-cylindrical. Every k-cylindrical invertible transformation of a Markov compact is a periodical transformation, the trajectories of which lie on the elements of the partition $\bigvee_{i=k}^{\infty} \eta_i$.

A homeomorphism T such that there exists a nowhere dense closed and invariant set with the property that in the complement of every open neighbourhood of this set the transformation T is a k-cylindrical map, where k depends on the neighbourhood, is called a transformation of the class $\mathcal{P}$. It means that the transformations of class $\mathcal{P}$ are limits of k-cylindrical (periodic) transformations in the sense of uniform convergence on every closed set belonging to the complement of any nowhere dense set.

We obtain an example of a transformation of class $\mathcal{P}$, which is of sufficient general form and to which we restrict our considerations, in the following way: We number the elements of the partition η_i by natural numbers from 1 to r_i $(i = 1,2,\dots)$. Using this order we can introduce a lexicographical ordering on the set of elements of $\bigvee_{i=1}^{k} \eta_i$ for all k, saying $(d_1,\dots,d_k) \succ (d_1',\dots,d_k')$ if $d_{k-i} = d_{k-i}'$ $(i = 0,1,\dots,i_0)$ and $d_{k-i_0} > d_{k-i_0}'$. On the space arises a particular order in the following form: sequences are compatible if their elements are equal from some place onward.

The ordering type of every class of comparable elements of $\mathfrak{X}$ is either

$\mathbb{Z}$ or $\mathbb{Z}_+$ or $\mathbb{Z}_-$ or $\overline{n,m}$. We call a Markov compact with ordering of the partitions minimal, if all of the classes with respect to this order, without two, are ordered by type $\mathbb{Z}$, one by type $\mathbb{Z}_+$ and one by type $\mathbb{Z}_-$. Minimality is a condition, which can be easily described by the matrices M_i:

STATEMENT 2: There exists a sequence $k_1 < k_2 < \dots$ such that the matrices $M_{k_i} \cdot M_{k_i+1} \cdot \dots \cdot M_{k_{i+1}-1}$ don't have zero-elements (i=1,2,...).

It is obvious that this statement does not depend on the special order on η . By variation of the numeration only the one-side orbits change.

DEFINITION 2: Let be given a Markov compact with ordered partitions η_i. A transformation T mapping the elements x onto an element x' which is greater than x in the sense of given order we call an adic transformation of this Markov compact (if x' does not exist, then T is not defined). If $\mathfrak{X}$ is minimal, we define the adic transformation uniting the halforbits $\mathbb{Z}_+$ and $\mathbb{Z}_-$.

STATEMENT 3: Adic transformation of a Markov compact belongs to the class $\mathcal{P}$, if the compact is minimal.

The proof follows directly by the definition of the order. In this case the exceptional set contains one and only one point. (The end of the orbit is of type $\mathbb{Z}_-$ and the beginning of the orbit of type $\mathbb{Z}_+$).

REMARK: A trajectorial partition of an adic transformation is the partition $\bigcap_{k=1}^{\infty} \bigvee_{i=k}^{\infty} \eta_i$ of $\mathfrak{X}$ (where $\cap$ denotes the set-theoretical intersection). Therefore the adic shift is the limit of cylindrical transformations, i.e. of permutations of the first coordinates of the sequence. We can imagine it as a tower of successive permutations of the elements of the partitions $\eta_1, \eta_2, \dots$. The adic shift is that model to which we reduce arbitrary ergodic automorphism of a Lebesgue space (see examples in § 4).

3. Basic theorem

THEOREM: Let (X,μ) be a Lebesgue space and S its ergodic automorphism with invariant measure. Let $\mathfrak{A}_0$ be a countable, under S invariant algebra of measurable sets in X. Then there exists a set $X_0 \subset X$, $X_0 \in \mathfrak{A}_0$ with $\mu(X_0) = 1$ and an automorphism $S' = S \bmod 0$, $S' X_0 = X_0$, and also a minimal Markov compact $\mathcal{X}$ (with ordered elements of partitions $(\eta_i)_{i \geq 1}$) such that $(X_0, S', \mathfrak{A}_0)$ and $(\mathcal{X}, T, \mathcal{B})$ are isomorphic, i.e. there exists a transformation t, defined on X_0: $X_0 \to \mathcal{X}$ with $t^{-1}\mathcal{B} = \mathfrak{A}_0$ ($\mathcal{B}$ - algebra of the cylinder sets) and $tS't^{-1} = T$, where T is an adic transformation.

Our task is to construct a sequence of partitions η_i on X with the necessary properties. But we begin with the construction of periodical approximations, which will correspond to cylindrical periodical transformations in the range.

Let B be an arbitrary measurable set in X with $m(B) > 0$, then every ergodic automorphism can be represented as an integrating one on the derivation S_B with respect to the subset B (see [11]). We fix the automorphism S and form the partitions $\zeta(B) = \zeta(B,S)$ and $\xi(B) = \xi(B,S)$ in the following way: $\zeta(B)$ consists of at most a countable number of elements $B_0 = B$, $B_1 = SB_0 \setminus B_0$, ... , $B_k = SB_{k-1} \setminus \bigcup_{i=0}^{k-1} B_i$. The elements of $\zeta(B)$ are ordered in a natural sense. The element $\xi(B)(x)$ containing the point $x \in B$ consists of all points of the form $S^i x$, where $S^i x \in B_i$. ζ and ξ are dual complementary. Let $r_B(x) = |\xi(B)(x)|$. Because S is ergodic it is $r_B(x) < \infty$ a.e. and we use the notion $r_{B,S} = \sup \operatorname{ess} r_B(x)$. We write $S^{(B)}x = Sx$, if $Sx \in \xi(B)(x)$, and $S^{(B)}x = S^{-r_B(x)}x$, if $Sx \notin \xi(B)(x)$. Then $S^{(B)}$ is a periodical automorphism with a period, depending on the point.

COROLLARY: Let $\mathfrak{A}_0$ be an everywhere dense algebra of measurable sets, invariant under the ergodic automorphism S. Then a sequence of sets $A_1 \supset A_2 \supset \ldots$ with $m(\bigcap_i A_i) = 0$ and $A_i \in \mathfrak{A}_0$ exists, such that $r_{A_i} < \infty$ for all $i = 1,2,\ldots$.

PROOF: We choose $A_1' \in \mathfrak{A}_0$ so that $\mu(A_1') < \frac{1}{2}$. We denote by A_1 the set $A_1 = A_1' \cup A_1''$, where A_1'' is the union of all, except of a finite number, elements of the partition $\zeta(A_1')$ with $m(A_1'') < \frac{1}{2} m(A_1')$. Since $\mathfrak{A}_0$ is invariant, it follows that $\zeta(A_1')$ is measurable with respect to $\mathfrak{A}_0$, that's why A_1'' and A_1 are elements of $\mathfrak{A}_0$. We remark that $r_{A_1,S} < \infty$. We consider $S_1 = S_{A_1}$, the derivation of S with respect to A_1, and $A_2' \subset A_1$ with $m(A_2') < \frac{1}{2} m(A_1')$. Now we apply to A_1, S_1 and A_2' the same method as to X, S, A_1' in the first step. In result of this we construct a set $A_2 \subset A_1$ with $r_{A_2,S_1} < \infty$ and therefore $r_{A_2,S} < \infty$. We have $m(A_1) < \frac{1}{2} + \frac{1}{4} = \frac{3}{4}$ and so we get $m(A_2) < \frac{3}{8} + \frac{3}{16} = (\frac{3}{4})^2$. Continuing this process we obtain a system of sets $A_1 \supset A_2 \supset \dots$ with $m(A_n) < (\frac{3}{4})^n$ and $r_{A_k,S} < \infty$, $A_n \in \mathfrak{A}_0$, q.e.d.

We denote $S^{(A_n)}$ by S_n. So the needed periodical automorphisms are constructed; it is evident, that $\lim_n m\{x \mid S_n x \neq Sx\} = 0$ holds. Now we construct a sequence of refinements $\zeta(A_n)$ which produce the structure of a Markov compact.

Let $\varrho_n \nearrow \varepsilon$ be an increasing sequence of finite partitions (here ε means the partition into points mod 0) consisting of elements of $\mathfrak{A}_0$ and generating $\mathfrak{A}_0$. We identify $X/\zeta(A_n)$ and A_n attaching to the element $\zeta(A_n)$ its single point from A_n. Let be $\pi_n: X/\zeta(A_n) \to A_n$. It is evident that $\pi_n \pi_{n-1} = \pi_n$ for all n, because $A_{n-1} \subset A_n$. Let be $\eta_1 = \zeta(A_n,S) \vee \varrho_1$. Then we define inductively the partitions η_k by

$$\eta_k = \pi_{k-1}^{-1}[\zeta(A_k,S_{A_{k-1}}) \vee \pi_{k-1}(\bigvee_{i=1}^{k-1} \eta_i \vee \varrho_k)] \qquad (k = 2,3,\dots).$$

The elements of η_k inherit the ordering of $\zeta(A_k,S_{A_{k-1}})$. We prove the fulfilment of the needed properties. Because

$$\zeta(A_k,S) = \zeta(A_1,S) \vee \pi_1^{-1}(\zeta(A_2,S_{A_1}))\dots\pi_{k-1}^{-1}(\zeta(A_k,S_{A_{k-1}})),$$

it holds $\zeta(A_k,S) \prec \bigvee_{i=1}^{k} \eta_i$ since the last partition is complementary to $\zeta(A_k,S)$, it means: to the trajectorial partition of the periodical automorphism S_k. Further $\varrho \prec \bigvee_{i=1}^{k} \eta_i$ holds, because $\varrho_k \prec \zeta(A_k,S) \vee \eta_k$.

Therefore $\bigvee_{i=1}^{\infty}\eta_i \succ \lim_k \varrho_k = \varepsilon$ is true. By construction it follows that the sequence $\{\eta_i\}$ is markovian, and that on one element D η_k the automorphism S_{k-1} has the same periods for all points, where it permutes the elements of $\bigvee_{i=1}^{k-1}\eta_i$ and lets the elements of $\bigvee_{i=k+1}^{\infty}\eta_i$ fixed. Finally all elements of all η_k are elements of $\mathfrak{A}_0$.

Now we define a Markov compact $\mathfrak{X}$ by the partitions $\{\eta_i\}$ and map X into the space $\prod_{k=1}^{\infty} X/_{\eta_k} = \tilde{\mathfrak{X}}$, i.e. into the countable product of the finite sets $X/_{\eta_k}$ with the weak topology. It is obvious that the mapping $\Theta : x \longrightarrow (\eta_1(x),\dots) \in \tilde{\mathfrak{X}}$ is a mod 0 - monomorphism and the closure of the range of the set with full measure is Markov compact $\mathfrak{X} \subset \tilde{\mathfrak{X}}$. The image of $\mathfrak{A}_0$ is the algebra of cylindrical sets. The Markov property of $\mathfrak{X}$ follows by the Markov property of $\{\eta_k\}_1^{\infty}$, where the elements of η_k are ordered. The transformation S is transferred to an adic shift, which follows by construction. Finally we prove the minimality of $\mathfrak{X}$. To show this we remark that by the ergodicity of S it follows, that for all elements $D \in \eta_k$ and $E \in \eta_s$ there exists a natural number $n = n(k,s)$ such that $m(S^n D \cap E) > 0$ holds. This shows that the elements of the matrix of intersections of η_k and η_s are positive (see statement 2, § 1). The theorem is proved.

REMARKS: 1) Obvious that method to construct the $\{\eta_k\}$ is not economical, because the refinement ζ_k is not accompanied by an extension of the trajectories of S_k. If possible $\bigvee_{i=1}^{k}\eta_i$ should be an 'almost' independent complement to ζ_k. The construction of economical realisations is a complicated task [see examples].

2) By the theorem [12] follows that there exists an invariant with respect to an ergodic automorphism algebra $\mathfrak{A}_0$ consisting of strict ergodic sets. If we use this algebra in the theorem then the corresponding adic transformation will be strict ergodic.

3) The asymptotic properties of the matrices of the intersections of the partitions η_k contain rich information for the automorphism about the metric invariants of the type of scale (see [6]).

4) The theorems are easily to generalize to nonergodic automorphisms. Thereby the Markov compact already will not be minimal.

5) The theorem is true for arbitrary countable groups of automorphisms, for which the Rochlin-Halmos-Lemma is fulfilled.

4. Examples

1. Let S be an ergodic automorphism with rational spectrum, i.e. with a spectrum, which is the union of finite subgroups of roots of unity of order $p_1, p_2, \dots$, where $p_{i+1}/p_i = r_i$. In this case the adic realisation is such one: the compact is $\mathcal{X} = \varprojlim \mathbb{Z}/p_i\mathbb{Z}$ and the transformation T is given by $Tx = x + 1$, where + denotes the addition in the additive group $\mathcal{X}$. As Markov compact $\mathcal{X}$ is $\prod_{i=1}^{\infty} \mathbb{Z}/r_i\mathbb{Z}$, whereby the matrices $M_i = \{m^i_{jk}\}$ are such that $m^i_{jk} \equiv 1$. If $r_i \equiv r$, then $\mathcal{X}$ is the additive group of entire r-adic numbers, what explains the term 'adic automorphism'.

2. Let be $S_\alpha x = (x + \alpha) \bmod 1$, where $x \in [0,1)$. If $\alpha \in [0,1)$ is irrational, then S_α is ergodic and its adic realisation can be represented in the following way: Let $\alpha = (n_1, n_2, \dots)$ be a partition of α into a continued fraction and $\mathcal{X} = \prod_{i=1}^{\infty} N_i$, $r_i = |N_i| = n_i + 1$. Let us consider matrices $M_i = \{m^i_{jk}\}$ of order $r_i \times r_{i+1}$ with $m^i_{jk} = 1$ if $j \geq 2$ and $m^i_{1k} = 0$ if $k < r_{i+1}$ and $m^i_{1r_{i+1}} = 1$ $(i = 1,2,\dots)$. This means

$$M_i = \begin{pmatrix} 0 & 0 & \dots & 0 & 1 \\ 1 & 1 & \dots & 1 & 1 \\ - & - & - & - & - \\ 1 & 1 & \dots & 1 & 1 \end{pmatrix} .$$

The Markov compact $\mathcal{X} \subset \tilde{\mathcal{X}}$ is constructed by $\{r_i\}$ and $\{M_i\}$. The adic shift on $\mathcal{X}$ is isomorphic to the shift S_α, more exactly: there exists an isomorphism of a measure space (on $\mathcal{X}$ exists one and only one invariant under the adic shift measure) transferring S_α into the adic shift. In this case the isomorphism is a Borel-isomorphism of topological spaces (circle and $\mathcal{X}$). This model is a version of the construction in [2,13], but there the construction is used for other aims.

Observing a proposal of the author M. and L. Handelsman introduced the isomorphism about which is spoken above.

Another adic realisation of the rotation of the circle is the following: Let be $\alpha_0 = \alpha$, $\alpha_i = 1 - \frac{1}{\alpha_{i-1}}$, $r_i = n_i = [\frac{1}{\alpha_{i-1}}] + 1$ $(i = 1,2,\dots)$, and let be $M_i = m^i_{jk}$ matrices of order $r_i \times r_{i+1}$, with $m^i_{jk} = 1$ for all j,k except $(1,r_{i+1})$ and $m_{1,r_{i+1}} = 0$, i.e.

$$M_i = \begin{pmatrix} 1 & \cdot & \cdot & \cdot & 1 & 0 \\ 1 & \cdot & \cdot & \cdot & 1 & 1 \\ - & - & - & - & - & - \\ 1 & \cdot & \cdot & \cdot & 1 & 1 \end{pmatrix} .$$

So every matrix M_i contains one and only one zero-element (see example 1). $\mathcal{X}(\Gamma(\{r_i\}, \{M_i\}))$ is the needed compact.

3. The shift of the torus and more than all automorphisms with continuous spectrum cause great difficulties. Here the algorithm of Perron and generalized partitions into continued fractions are useful. It is interesting to put the inverse tasks, i.e. to study the ergodic properties of an adic shift on given Markov compacts and on graphs. The graph of Pascal and the graph of Jung generate examples of shifts, about which nothing is known.

REFERENCES

[1] Veršik, A.M.: Uniform algebraic approximation for shift and product operators, Dokl. AN SSSR 259/3 (1981), 526-529.

[2] Pimsner, M. and D. Voiculescu: Imbedding the irrational rotation C*- algebra into an AF-algebra. Prepared in: Journ. of Operator Theory (1981).

[3] Veršik, A.M.: Countable groups, near to finite ones. In: Greenleaf, F.P.: Topological groups with invariant mean. Moskov 1972.

[4] Veršik, A.M.: Does there exist uniform algebraic approximation for product and shift operators? Publ.sci.sem. LOMI 81 (1978), 77-81.

[5] Rieffel, M.A.: Irrational rotation C* - algebras. In: Internat. Congr. of Math. Helsinki 1978.

[6] Veršik, A.M.: Four definitions of the scale of an automorphism. Funct. Anal. and Appl. 7/3 (1973), 1-17.

[7] Katok, A.B., Ya.G.Sinai and A.M.Stepin: Theory of dynamical systems and of general groups of transformations with invariant measure. In: Results of Sci., Math. Anal. Moskov 1975, 129-261.

[8] Veršik, A.M.: Multivalued maps with invariant measure (polymorphisms) and Markov operators. Publ.Sci.Sem. LOMI 72 (1977), 26-61.

[9] Veršik, A.M. and O.A.Ladyženskaja: On evolution of measures, defined by Navier-Stokes-equation, and on solubility of the Cauchy problem for the statistical equation of E. Hopf. Publ.Sci.Sem. LOMI 59 (1976), 3-24.

[10] Effros, E.G.: The dimension group. Preprint.

[11] Kornfel', I.P., Ya.G.Sinai and S.V.Fomin: Ergodic Theory. Moskov 1980.

[12] Hansel, G. and J. Raoult: Ergodicity, uniformity and unique ergodicity. Indiana Univ. Math. J. 23/3 (1973), 221-237.

[13] Effros, E.G. and C.L.Shen: Approximately finite C^*- algebras and continued fractions. Preprint.

ON CODING OF MARKOV SOURCES

R. J. Grigorchuk[1], A. M. Stepin[2]

For Markov shifts in terms of generating functions necessary and sufficient conditions of the coding based on Dehn's algorithm are found. The equations for the generating functions of the code without overlapping are deduced. Amenability condition for semigroup associated with a code is given. The variational characterization of the efficiency of hierarchical coding is obtained. This characterization is dual to the well known variational principle for topological entropy. There is proved the existence of a source which is effective coded by the given hierarchical code.

1. Basic definitions and formulation of the problem

Coding of Markov shifts with equal entropy, done by Ornstein, possesses infinite reach and infinite memory. Recently Keane and Smorodinsky [5] showed the existence of finitary coding (i.e. with finite a.s. reach and finite a.s. memory) of Markov shifts with equal entropy. In this connexion it is interesting to get constructive procedures for finitary coding. As an example we regard a procedure for hierarchical coding.

Let A be a finite alphabet. We denote by Ω the space of all twosided infinite sequences $\omega = \{\omega_i\}_{-\infty}^{\infty}$ of symbols out of A. The pair (Ω , μ), where μ is a Borel probability measure on Ω , is the mathematical description of an information source and the elements of the space Ω are considered as messages of this source. The coding problem of the messages is to construct a measurable map (the code) from the space Ω into another sequence space Ω', defined on a set of full measure. A code φ is called invertible iff it is an isomorphism of the

---

[1] Moscow Institute of Engineers for Railway Transport, 10 30 55 Moscow, ul. Obraszova 15

[2] Moscow State Lomonosov University, 117234 Moscow, Lenin mountains

shifts to the left T_{Ω} and $T_{\Omega'}$ acting on the spaces (Ω, μ) and $(\Omega', \varphi_* \mu)$ respectively. A code is a stationary code iff $\varphi T_{\Omega} = T_{\Omega'}\varphi$. We call a stationary code φ hierarchical, iff for all sequences $\omega \in \Omega$ a neighbourhood N_ω of the form $\{\eta \in \Omega : \eta_k = \omega_k , m(\omega) \leq k \leq n(\omega)\}$ exists such that the symbols $[\varphi(\eta)]_k$, $m(\omega) \leq k \leq n(\omega)$, do not depend on $\eta \in N_\omega$.

It is possible to give a constructive description of the procedure of the hierarchical coding. For this purpose we introduce necessary definitions. Let the alphabet $A = \{a_i\}_1^r$ be fixed. The length of a word W in the alphabet A we denote by $|W|$. $|W|_i$ (resp. $|W|_{ij}$) is the sign of the occurance of the symbol $a_i \in A$ (resp. of the subword $a_i a_j$) in the word W. The product of the words V_1 and V_2 is denoted by $V_1 V_2$. The insertion of the word U into the word V is the word $V_1 U V_2$ where V_1 and V_2 are nonempty words with $V_1 V_2 = V$. If $W = a_{i_1} a_{i_2} \dots a_{i_n}$ is a word, the word $\tau(W) = a_{i_2} a_{i_3} \dots a_{i_n} a_{i_1}$ is the cyclic shift of W.

We fix a collection $K = \{W_j\}$ of words in the alphabet A and denote by $\mathcal{E}(K)$ the closure of K with respect to the multiplication and insertion operations. In the following we are interested in the calculation or the estimation of the number of words with length n contained in $\mathcal{E}(K)$. We remark that in many cases it can be decided wether a given word is in $\mathcal{E}(K)$ or not by an algorithm analog to that given by Dehn for the problem of the equality of words (see [8]).

A word $W \in \mathcal{E}(K)$ is called simple, iff there is no representation of W in form of a product of two nonempty words in $\mathcal{E}(K)$. If $W = a_{i_0} a_{i_1} \dots a_{i_n}$ is a word in the alphabet A then let $\overline{W}$ be the cylinder set

$$\{\omega : \omega_0 = a_{i_0}, \omega_1 = a_{i_1}, \dots, \omega_n = a_{i_n}\} \subset \Omega = A^{\mathbb{Z}} .$$

If we enumerate the words of the set $\mathcal{E}(K)$ we get the sequence $\{B_j\}$. Let us consider the set

$$\Omega_k = \bigcap_{n=1}^{\infty} \bigcup \{T^s \overline{B}_j , j \geq n , 1 \leq s \leq |B_j| \} .$$

It consists of the sequences ω possessing the following property: for all integers M and N, $M < N$, there exist integers $m < M$ and $n > N$

such that the word $\omega_m \dots \omega_n$ is in $\mathcal{E}(K)$ and hence is invariant under the shift operations.

We assume that the collection K' of words in the alphabet A' and the map $\varphi: K \longrightarrow K'$ are such that there exists an extension $\varphi_1: \mathcal{E}(K) \longrightarrow \mathcal{E}(K')$ of the map φ, possessing the properties

1) $|W| = |\varphi_1(W)|$, $W \in \mathcal{E}(K)$
2) $\varphi_1(UV) = \varphi_1(U)\varphi_1(V)$, $U, V \in \mathcal{E}(K)$
3) if $U = U_1U_2 \in \mathcal{E}(K)$, $W \in \mathcal{E}(K)$ then
   $\varphi_1(U_1WU_2) = U_1'\varphi_1(W)U_2'$ where $U_1'U_2' = \varphi_1(U)$ and $|U_1| = |U_1'|$
4) in every case, if the last k symbols of the word $U \in \mathcal{E}(K)$ are equal to the first k symbols of the word $V \in \mathcal{E}(K)$, the words $\varphi_1(U)$ and $\varphi_1(V)$ have this property.

Then the map φ_1 can be extended to a stationary map $\varphi_2: \Omega_K \longrightarrow \Omega_{K'}$. If $\mu(\Omega_K) = 1$ then φ_2 is a hierarchical code. It is not difficult to show that every hierarchical code can be obtained in this way.

If the map $\varphi: K \longrightarrow K'$ is bijective then the corresponding code φ_2 is invertible.

Because the invertible hierarchical code essentially is uniquely defined by the bijective map $K \longrightarrow K'$ we will call (allowing some freedom) the collection K a code. If $\mu(\Omega_K) = 1$ we will say that the code K covers the measure μ.

Let us consider two cases, when one is able to express the existence of the extension φ_1 fulfilling the conditions 1) - 4) in terms of the coding collections K and K'. A code possessing the properties

i) no word of K is a part of another word in K,

ii) the beginning of any word is not the end of any another word

is called S-code or a code without overlapping.

<u>Examples of S-codes:</u>

a) The canonical code: $A = \{a_i\}_0^r$,

$$K = \{\underbrace{a_0 \dots a_0}_{l_1} a_1, \underbrace{a_0 \dots a_0}_{l_2} a_2, \dots , \underbrace{a_0 \dots a_0}_{l_r} a_r \}, \; l_i > 0, i=1, \dots, r.$$

b) The alphabet $A = \{\alpha_i,\ i=1,2,\dots,r\ ;\ \beta_j,\ j=1,2,\dots,s\}$
$K = \{\alpha_i\beta_j\ ,\ i=1,2,\dots,r;\ j=1,2,\dots,s\}$.

For a given S-code $K = \{W_j\}_1^n$ we construct the canonical S-code $K_o = \{V_j\}_1^r$ with the alphabet $A_o = \{a_i\}$ by setting $V_j = \underbrace{a_o\cdots a_o}_{l_j}a_j$, $l_j = |W_j| - 1$.

The map φ: $K \to K_o$, $W_j \to V_j$ is the most simple example where the conditions 1) - 4) are fulfilled and hence defines a hierarchical coding.

S-codes and the procedure of coding, based on their exploitation, have been examined in the papers of Mechalkin[9], Zaslavsky[4], Lewenschtein[6], Lifschitz[7] and other authors.

Let $\tilde{K}$ be the collection of words obtained from K by the addition of all cyclic shifts of the words $W \in K$. Let K' be a collection of words in the alphabet A' and the map φ: $K \to K'$ be such that the natural extension $\tilde{\varphi}$: $\tilde{K} \to \tilde{K}'$ possesses the following properties:

α) $|W| = |\tilde{\varphi}(W)|$, $W \in \tilde{K}$

β) in all cases, when the last k symbols of the word $U \in \tilde{K}$ are equal to the first k symbols of the word $V \in \tilde{K}$, then the words $\tilde{\varphi}(U)$ and $\tilde{\varphi}(V)$ have this property, too.

Then one can extend the map φ: $K \to K'$ to a map $\mathcal{E}(K) \to \mathcal{E}(K')$ possessing the properties 1) - 4) and hence to a stationary coding $\Omega_K \to \Omega_{K'}$.

<u>Example</u>. Let $A = \{a,b,c,d\}$, $K = \{abc\ ,\ cd\}$, $A' = \{w_o,w_1,w_2\}$, $K' = \{w_1w_ow_o\ ,\ w_ow_2\}$. The map φ: $K \to K'$ with $\varphi(abc)=w_1w_ow_o$, $\varphi(cd)=w_ow_2$ fulfils the conditions α) and β).

It follows from the above that the problem of hierarchical coding leads to the following problems:

a) Given a source (Ω, μ) and a code K in the alphabet of the source, what are the conditions for $\mu(\Omega_K) = 1$?

b) Given a code K in the alphabet A and an 'alphabet of the interpreter' A', when does there exist a code K' and a map $\mathcal{E}(K) \to \mathcal{E}(K')$ with the properties 1) - 4) ?

In this paper we consider the first of these questions.

2. Necessary conditions for coding

We adjoin to the code K the following power series:

$$\Phi_B(x_1,\dots,x_r) = 1 + \sum C_{i_1\dots i_r} x_1^{i_1}\dots x_r^{i_r} \qquad (1)$$

where $C_{i_1\dots i_r}$ is the number of words $W \in \mathcal{E}(K)$ such that $|W|_1 = i_1$, ..., $|W|_r = i_r$; and

$$\Phi_M^{k,l}(x_{11},\dots,x_{rr}) = \sum C^{k,l}_{\alpha_{11}\dots\alpha_{rr}} x_{11}^{\alpha_{11}}\dots x_{rr}^{\alpha_{rr}}$$

where $C^{k,l}_{\alpha_{11}\dots\alpha_{rr}}$ is the number of words $W \in \mathcal{E}(K)$ beginning with the symbol a_k and ending with the symbol a_l such that $|W|_{ij} = \alpha_{ij}$, $i,j = 1,\dots,r$; and

$$\Phi_M = 1 + \sum_{k,l} \Phi_M^{k,l} .$$

We call the series Φ_B and Φ_M the counting series of the code K. They are connected by the equation

$$\Phi_B(x_1^2,\dots x_r^2) = 1 + \sum_{k,l} x_k x_l \, \Phi_M^{k,l}(x_1^2,\dots,x_i x_j,\dots,x_r^2) .$$

If the code coincides with the alphabet $A = \{a_i\}$, i.e. $W_i = a_i$, $i = 1,\dots,r$, then

$$\Phi_B = (1 - x_1 - \dots - x_r)^{-1} .$$

Because the numbers $C_{i_1\dots i_r}$ in (1) for any code K are not greater than

$$\frac{(i_1 + \dots + i_r)!}{i_1!i_2!\dots i_r!} ,$$

the series Φ_B converges in any multicylinder $|x_1| < p_1,\dots,|x_r| < p_r$, $p_1 + p_2 + \dots + p_r = 1$.

In an analogous way it is possible to show, calculating the function Φ_M for the case when K coincides with the alphabet, that for any K the series Φ_M converges in any multicylinder $|x_{ij}| < p_{ij}$ where $\{p_{ij}\}$ is a stochastic matrix.

<u>Theorem 1</u>. Let μ be a stationary Markov measure with transition probabilities $\{p_{ij}\}$. If the code K is choosen so, that $\mu(\Omega_K) = 1$, then the point $(p_{11},\dots,p_{ij},\dots,p_{rr})$ is a singular point for the function

Φ_M and

$$\sum_{i,j} p_{ij} \frac{\partial \Phi_M}{\partial x_{ij}}(p_{11},\dots,p_{rr}) = \infty .$$

Proof. Because of $\mu(\Omega_K) = 1$ and $\Omega_K = \bigcap_{n=1}^{\infty} \bigcup_{j \geq n} E_j$ with $E_j = \bigcup\{T^s\overline{B_j} , 1 \leq s \leq |B_j|\}$, one obtains by the lemma of Borel-Cantelli

$$\sum_{j=1}^{\infty} |B_j| \mu(\overline{B_j}) = \sum_{j=1}^{\infty} \sum_{1 \leq s \leq |B_j|} \mu(T^s\overline{B_j}) \geq \sum_{j=1} \mu(E_j) = \infty .$$

By the definition of the function Φ_M follows that

$$\sum_{i,j} p_{ij} \frac{\partial \Phi_M}{\partial x_{ij}}(p_{11},\dots,p_{rr}) = \sum_{l} (|B_l| - 1) \mu(\overline{B_l})$$

and for this reason

$$\sum_{|\overline{B_l}| \geq 2} |B_l| \mu(\overline{B_l}) \leq 2 \sum_{i,j} p_{ij} \frac{\partial \Phi_M}{\partial x_{ij}}(p_{11},\dots,p_{rr}).$$

For the rest of the proof one uses a multidimensional analogon of the theorem of Hadamard-Pringsheim.

For a Bernoulli measure μ with onedimensional distribution $(p_i)_1^r$ a necessary condition for $\mu(\Omega_K) = 1$ is: the point $(p_1,\dots,p_r)$ is a singular point for Φ_B where the series $\sum p_i \, {}^{\partial \Phi_B}/_{\partial x_i}(p_1,\dots,p_r)$ diverges.

Example. The code $\{a_1a_2a_2 , a_1a_1a_2\}$ does not cover any Bernoulli measure.

We refer to one algebraic application of theorem 1. Let $K = \{W_i\}$ be a code in the alphabet $A = \{a_i\}_1^r$ such that there exists a Bernoulli measure with $\mu(\Omega_K) = 1$. Then the semigroup π_K , given by the generating elements $a_1,\dots,a_r$ and by $W_j = e$, and permitting a right division is amenable, i.e. possesses a right invariant finite additive measure (see [3]). For the proof of this fact we consider a random walk on π_K with transition probabilities $p(e,a_i) = p_i$. According to our assumption the point $x = 1$ is a singular point for the function $\Phi_B(p_1x,\dots,p_rx)$ that can be taken as a generating function for the probabilities $p^{(n)}$ of the recurrence to the unit of the semigroup π_K after n steps. That's why the spectral radius R of the walk on π_K

(i.e. $\overline{\lim_{n\to\infty}} \{p^{(n)}\}^{1/n}$) is equal to 1. But because R never exceeds the the $l_2(\Pi_K)$-norm of the transition operator of the walk [11], the amenability of Π_K follows now by the criterion of Day [3] for the amenability of a semigroup with division.

Example. Let $K = \{\underbrace{a_0\ldots a_0}_{l_j} a_j\}_1^r$ be a canonical code. The semigroup

$$\Pi_K = \langle a_0, a_1, \ldots a_r \mid \underbrace{a_0\ldots a_0}_{l_1} a_1 = e \ , \ \ldots \ , \ \underbrace{a_0\ldots a_0}_{l_r} a_r = e \rangle$$

permits a right division. This follows by the fact that every element $g \in \dot{\Pi}_K$, thought as a class of equivalent words, possesses a unique nondivisible representative (see for ex [1]). It is well known [7] that for any $l_1,\ldots,l_r$ there exists a Bernoulli measure covered by the code K. For this reason the semigroup Π_K is amenable for any $l_1,\ldots,l_r$.

3. Equations for the generating functions Φ_B and Φ_M

We will see that in the case of an S-code theorem 1 is reversible. To prove this fact we introduce equations fulfilled by the generating functions Φ_B and Φ_M.

Theorem 2. The counting function Φ_B of an S-code fulfils the equation

$$1 - \Phi_B + \sum_{W\in K} x_1^{|W|_1} \ldots x_r^{|W|_r} \Phi_B^{|W|} = 0 \ .$$

Proof. The products of coding words are called words of first kind. The word $U \in \mathcal{E}(K)$ is called a word of (n+1)-kind iff we get a word of kind n if we remove from the original word all coding subwords. (We remark that every word in $\mathcal{E}(K)$ contains by definition any subword). We show that the counting functions $\Phi_n(x_1,\ldots,x_r)$ of words of kind not greater than n fulfil the recurrence expression

$$\Phi_{n+1}(x_1,\ldots,x_r) = [1 - \sum_{W\in K} x_1^{|W|_1}\ldots x_r^{|W|_r} \Phi_n^{|W|-1}(x_1,\ldots,x_r)]^{-1} \quad (2)$$

For this purpose we write the word U of kind not greater than (n+1) in form of a product of simple words. The simple word V of kind not greater than (n+1) has the form

$$V = a_{i_1} V_1 a_{i_2} V_2 \dots V_{s-1} a_{i_s} \tag{3}$$

where $W = a_{i_1} a_{i_2} \dots a_{i_s} \in K$ and $V_1, \dots, V_{s-1}$ are words of $\mathcal{E}(K)$ of kind not greater than n. The counting function of the words of the form (3) is equal to

$$x_1^{|W|_1} \dots x_r^{|W|_r} \Phi_n^{|W|-1}(x_1,\dots,x_r) \, .$$

The counting function of the words of kind not greater than (n+1) that are products of l simple words from $\mathcal{E}(K)$ is equal to

$$\Big[\sum_{W \in K} x_1^{|W|_1} \dots x_r^{|W|_r} \;{}_n^{|W|-1}(x_1,\dots,x_r) \Big]^l \, .$$

If we sum up these terms with respect to n, we get (2). The coefficients of the series are majorized by the coefficients of the power series

$$\sum \frac{(i_1+i_2+\dots+i_r)!}{i_1!\,i_2!\dots i_r!} x_1^{i_1} \dots x_r^{i_r} = \frac{1}{1 - x_1 - \dots - x_r} \, .$$

For this reason the series Φ_n converge in every multicylinder $|x_1| < p_1, \dots, |x_r| < p_r$, $\sum p_i = 1$. On the other hand the coefficients of the series Φ_{n+1} majorize the corresponding coefficients of the power series Φ_n and are stable for $n \to \infty$. It follows that the sequence of the functions Φ_n converges in every point of the multicylinder $|x_1| < p_1, \dots, |x_r| < p_r$. Hence one can go in equation (2) to the limit $n \to \infty$. In the limit we obtain that the counting function Φ_B fulfils the equation

$$\Phi_B = \Big[1 - \sum_{W \in K} x_1^{|W|_1} \dots x_r^{|W|_r} \Phi_B^{|W|-1} \Big]^{-1} \, .$$

The theorem is proved.

To get equations for the $\Phi_M^{k,l}$ we introduce auxiliary functions

$$\sigma_W = x_{11}^{i_{11}} \dots x_{rr}^{i_{rr}} \, , \quad i_{11} = |W|_{11} \, , \dots , \; i_{rr} = |W|_{rr} \, .$$

Let us decompose the alphabet A into parts B, C and D such that every word $W \in K$ begins with an element of B and ends with an element of C. (The subsets B, C and D are disjoint because K is a code without overlapping). We denote $L_{b,c}(X) = \sum \sigma_W$ where the summation is over all $W \in K$, which begin with the symbol $b \in B$ and end with the symbol $c \in C$

and $X = (x_{11},\dots,x_{rr})$. Let us introduce the matrix

$$T = \begin{pmatrix} 0 & L(X) & 0 \\ X' & 0 & 0 \\ 0 & 0 & 0 \end{pmatrix}$$

where X' is the matrix with elements x_{cb} , $c \in C$, $b \in B$ and $L(X)$ is the matrix with elements $L_{b,c}(X)$. We write $\Gamma_{ij}(L(X) , X')$ for the elements of T.

Theorem 3. For every S-code K the functions $\Phi_M^{k,l}$ fulfil the system of equations

$$\Phi_M^{k,l} = \Gamma_{k,l}(L(Y) , X') \tag{4}$$

where Y is the matrix with elements

$$y_{ij} = (x_{ij} + \sum_{\alpha,\beta=1}^{r} x_{i\alpha} x_{\beta j} \Phi_M^{\alpha,\beta})$$

and k,l are such that $a_k \in B$, $a_l \in C$.

We remark that in case of a finite S-code the $\Phi_M^{k,l}$ are algebraic functions because the Γ_{ij} are rational functions.

Proof. We consider the power series $\Phi_n^{k,l} = \sum x_{11}^{|W|_{11}} \dots x_{rr}^{|W|_{rr}}$ where the sum is over all words with length not greater than n and beginning with the symbol a_k and ending with the symbol a_l.

Let us derive recurrent relations for $\Phi_n^{k,l}$. To do this we consider the matrix

$$T_n = \begin{pmatrix} 0 & L^{(n)} & 0 \\ X' & 0 & 0 \\ 0 & 0 & 0 \end{pmatrix}$$

where the matrix element $L_{\alpha,\beta}^{(n)}$ has the form

$$\sum \prod_{i,j} (x_{ij} + \sum_{k,l} x_{ik} x_{lj} \Phi_n^{k,l})^{|W|_{ij}}$$

and the outer sum runs over all words $W \in K$, beginning with a_α and ending with a_β . The function $\Phi_{n+1}^{k,l}$ turns out to be equal to the (k,l)-th element of the matrix $(1 - \tilde{T}_n)^{-1}$. To prove this we represent every word U with length not greater than n+1 in the form of a product of simple words. A simple word with length not greater than n+1 beginning with the symbol a_k and ending with the symbol a_l has the form

$$U = a_k U_1 a_{i_1} U_2 \dots U_{s-1} a_{i_s} U_s a_l \qquad (5)$$

where $W = a_k a_{i_1} \dots a_{i_s} a_l \in K$ and $U_1, \dots, U_s$ are words of $\mathcal{E}(K)$ with length not greater than n. The generating function of the words of form (5) is equal to $L^{(n)}_{k,l}$. Like the (k,l)-th matrix element of the m-th power of the stochastic matrix which gives the probability of the transition from the k-th state to the l-th state after m steps, the (k,l)-th matrix element of the m-th power of the matrix $\tilde{T}_n$ gives a term of the generating function of the words with length not greater than n , with first symbol a_k and last symbol a_l and representable as a product of m simple words of $\mathcal{E}(K)$.

The coefficients of the series $\Phi^{k,l}_n$ are not greater than the corresponding coefficients of the series $\Phi^{k,l}_M$ and consequently they converge in every polycylinder $|x_{ij}| < p_{ij}$, $\sum p_{ij} = 1$, i=1,...,r. Moreover, the coefficients of series Φ_{n+1} are greater or equal than the coefficients of the series Φ_n and stabilize if $n \to \infty$. Therefore the function sequence $\Phi^{k,l}_n$ converges to $\Phi^{k,l}_M$ at the polycylinder $|x_{ij}| < p_{ij}$. Consequently we can take the limit of

$$\Phi^{k,l}_{n+1} = [(1 - \tilde{T}_n)^{-1}]_{k,l}$$

for $n \to \infty$. For the limit we get, that the generating functions $\Phi^{k,l}_M$ fulfil the system (4).

<u>Example</u>. We consider a code K in the alphabet $A = \{a_i\}_1^r$ which consists of the words with a_1 as the first and one of $a_2, \dots, a_r$ as the last symbol. In this case the system of equations (4) has the form

$$\Phi^{1,l}_M = \sum_{\substack{W \in K \\ W = W' a_l}} \varrho(W) \left[1 - \sum_{j=2}^{r} x_{j1} \sum_{\substack{W \in K \\ W = W' a_j}} \varrho(W) \right]^{-1} , \quad i=2,\dots,r ,$$

where

$$\varrho(W) = \prod_{k=1}^{|W|-1} \left(x_{i_k} x_{i_{k+1}} + \sum_{\alpha=2}^{r} x_{i_k 1} x_{\alpha i_{k+1}} \Phi^{1,\alpha}_M \right)$$

if $W = a_{i_1} a_{i_2} \dots a_{i_{|W|}}$.

In the special case of an alphabet $A = \{a_1, a_2\}$ and the code $\{a_1 a_2\}$ we have

$$\Phi_M^{1,2} = \frac{x_{12} + x_{11}x_{22}\Phi_M^{1,2}}{1 - x_{21}(x_{12}+x_{11}x_{22}\Phi_M^{1,2})} \quad .$$

Let us consider the generating functions

$$\Psi_B(x_1,\dots,x_r) = \sum d_{i_1\dots i_r} x_1^{i_1} \dots x_r^{i_r}$$

where $d_{i_1\dots i_r}$ is the number of simple words $W \in \mathcal{E}(K)$ with $|W|_1 = i_1$, $\dots$, $|W|_r = i_r$,

$$\Psi_M^{k,l}(x_{11},\dots,x_{rr}) = \sum d^{k,l}_{\alpha_{11}\dots\alpha_{rr}} x_{11}^{\alpha_{11}} \dots x_{rr}^{\alpha_{rr}}$$

where $d^{k,l}_{\alpha_{11}\dots\alpha_{rr}}$ is the number of simple words $W \in \mathcal{E}(K)$ with a_k as first and a_l as last symbol and $|W|_{ij} = \alpha_{ij}$, $i,j = 1,\dots,r$,

$$\Psi_M = \sum \Psi_M^{k,l} \ .$$

The generating function Ψ_B of simple words is connected with the generating function Φ_B by the relation

$$\Phi_B = (1 - \Psi_B)^{-1} \quad . \tag{6}$$

<u>Proposition 1</u>. The series

$$\sum_{i,j} p_{ij}\frac{\partial \Phi_M}{\partial x_{ij}} \quad , \quad \sum_{i,j} p_{ij}\frac{\partial \Psi_M}{\partial x_{ij}}$$

either both converge or both diverge at the point $(p_{11},\dots,p_{rr})$, supposed that $p_{ij} > 0$, $i,j = 1,\dots,r$.

Proof. By $\tilde{\Phi}_M$ (resp. $\tilde{\Psi}_M$) we denote the $r \times r$-matrix with the (k,l)-th element $\Phi_M^{k,l}$ (resp. $\Psi_M^{k,l}$), if $a_k \in B$, $a_l \in C$, and zero otherwise. We will prove that $\tilde{\Phi}_M = \tilde{\Psi}_M(1 - X\tilde{\Psi}_M)^{-1}$ with

$$X = \begin{pmatrix} 0 & 0 & 0 \\ X' & 0 & 0 \\ 0 & 0 & 0 \end{pmatrix} \quad .$$

(We remark that this relation is a Markovian analogy of the relation (6)). Really

$$\tilde{\Phi}_M = \tilde{\Psi}_M + \tilde{\Psi}_M X \tilde{\Psi}_M + \tilde{\Psi}_M X \tilde{\Psi}_M X \tilde{\Psi}_M + \dots = \tilde{\Psi}_M(1 - X\tilde{\Psi}_M)^{-1}, \tag{7}$$

consequently

$$\frac{\partial \tilde{\Phi}_M}{\partial x_{ij}} = \frac{\partial \tilde{\tilde{\Phi}}_M}{\partial x_{ij}} (1-X\tilde{\Psi}_M)^{-1} + \tilde{\Psi}_M(1-X\tilde{\Psi}_M)^{-1}(\frac{\partial X}{\partial x_{ij}} \tilde{\Psi}_M + X\frac{\partial \tilde{\Psi}_M}{\partial x_{ij}})(1-X\tilde{\Psi}_M)^{-1}. \quad (8)$$

We suppose that $\sum p_{ij} \frac{\partial \tilde{\Psi}_M}{\partial x_{ij}} < \infty$ at the point $(p_{11},\ldots,p_{rr})$. Because of $p_{ij} > 0$ all elements of the matrices $\tilde{\Psi}_M(p_{ij})$ and $\frac{\partial \tilde{\Psi}_M}{\partial x_{kl}}(p_{ij})$ are finite.

If the series $\sum p_{ij}\frac{\partial \tilde{\Phi}_M}{\partial x_{ij}}(p_{11},\ldots,p_{rr})$ would diverge, there would be pairs (i,j) and (k,l) with partial derivation $\partial \tilde{\Phi}_M^{k,l}/\partial x_{ij}$ infinite at the point $(p_{11},\ldots,p_{rr})$. From (8) follows, that $\det(1 - X\tilde{\Psi}_M) = 0$ at the point (p_{ij}) and therefore, in accordance with (7) at least one of the elements of the matrix $\tilde{\tilde{\Phi}}_M$ is infinite at $(p_{11},\ldots,p_{rr})$. We will prove, however, with help of theorem 3, that all elements of $\tilde{\tilde{\Phi}}_M$ are finite. Really, all nonzero matrix elements are solutions of the system of equations

$$\tilde{\Phi}_M^{k,l}(x_{11},\ldots,x_{rr}) = \sum_{\{\alpha_{ij}\}} P_{\{\alpha_{ij}\}}(x_{11},\ldots,x_{rr}) \prod_{i,j} (\tilde{\Phi}_M^{ij})^{\alpha_{ij}} \quad (9)$$

where $P_{\{\alpha_{ij}\}}$ is a polynomial with positive coefficients. Now, let us suppose that there exists a pair (k,l) such that

$$\tilde{\Phi}_M^{k,l}(p_{11}t,\ldots,p_{rr}t) \longrightarrow \infty$$

for $t \longrightarrow 1$. Because at the right side of the equation (9) we have $\tilde{\Phi}_M^{k,l}$ to a power not less than 2 for $t \longrightarrow 1$ the right side of (9) grows faster than the left one. The resulting contradiction proves the proposition.

4. Conversion of theorem 1

To convert theorem 1 in the case of S-codes we need the following combinatorial fact.

Lemma 1. If K is a code without overlapping and the beginning of the simple word $U_1 \in \mathcal{E}(K)$ is the end of the simple word $U_2 \in \mathcal{E}(K)$, then $U_1 = U_2$.

The proof we leave to the reader.

<u>Corollary</u>. If U_1 and U_2 are simple words of $\mathcal{E}(K)$ and neither U_1 is subword of U_2 , nor U_2 a subword of U_1 , then

$$T^{i_1}\overline{U}_1 \cap T^{i_2}U_2 = \emptyset , \quad 0 \leq i_1 < |U_1| , \quad 0 \leq i_2 < |U_2| .$$

<u>Definition</u>. Let us order the simple words of $\mathcal{E}(K)$ into a sequence $\{C_j\}$. We will say, that the code K connectedly covers the measure μ , if the set $\Omega'_K = \bigcap_{n=1}^{\infty} \bigcup_{j \geq n} \bigcup_{s=1}^{|C_j|} T^s\overline{C_j}$ has full measure.

The set Ω'_K is invariant with respect to shift transformations and consists of the sequences ω having the property: For all integers M , N , $M < N$, one can find integers $m < M$, $n > N$ such that the word $\omega_m \dots \omega_n$ is a simple word of $\mathcal{E}(K)$. The elements of the set Ω'_K are called simple realizations.

<u>Theorem 4</u>. If the stochastic matrix $\{p_{ij}\}$ is ergodic and there is an S-code such that the point $(p_{11},\dots,p_{rr})$ is singular for the function Φ_M and it holds

$$\sum p_{ij} \frac{\partial \Phi_M}{\partial x_{ij}}(p_{11},\dots,p_{rr}) = \infty ,$$

then the code K covers the stationary Markov measure μ with transition probabilities $\{p_{ij}\}$.

Proof. First we will prove that the set Ω'_K of the simple realizations has a representation in the following form

$$\bigcap_{n=1}^{\infty} \bigcup_{j \geq n} G_j$$

where $G_j = T^j [\bigcup_{|C_s| \geq j} \overline{C_s}]$ and C_s - simple words of $\mathcal{E}(K)$.

Let $\omega \in \bigcup_{j \geq n} G_j$ for every n , then there exists an increasing index sequence n_1 , n_2 , ... such that $\omega \in G_{n_1}$, $\omega \in G_{n_2}$, ... Therefore we find a sequence $\{s_k\}$ such that $\omega \in T^{n_k}\overline{C_{s_k}}$, $k=1,2,\dots$.
In accordance with the corollary of Lemma 1 we have that C_{s_k} is a subword of the word $C_{s_{k+1}}$. Therefore the sequence ω is a simple realization. The converse is obvious.
We make use of a contemporary version of the Lemma of Borel-Cantelli ([10]): if G_k is a set sequence with $\sum\mu(G_k) = \infty$ and such that for

some $\gamma > 0$

$$\underline{\lim}_{n\to\infty}\Big[\sum_{k=1}^{n}\mu(G_k)\Big]^{-2}\sum_{k,l=1}^{n}\mu(G_k\cap G_l) < \gamma ,$$

then

$$\mu\Big[\bigcap_{n=1}^{\infty}\bigcup_{j\geq n} G_j\Big] \geq \frac{1}{\gamma} .$$

In our case

$$\sum_{j=1}^{\infty}\mu(G_j) = \sum_{j=1}^{\infty}|C_j|\mu(\overline{C_j}) .$$

In consequence of proposition 1 the divergence of the series

$$\sum_{i,j} p_{ij}\frac{\partial\Phi_M}{\partial x_{ij}}(p_{11},\dots,p_{rr})$$

involves the divergence of the series

$$\sum_{i,j} p_{ij}\frac{\partial\Psi_M}{\partial x_{ij}}(p_{11},\dots,p_{rr}) .$$

From the definition of the function Ψ_M we obtain that

$$\sum_{i,j} p_{ij}\frac{\partial\Psi_M}{\partial x_{ij}}(p_{11},\dots,p_{rr}) = \sum_{s=1}^{\infty}(|C_s| - 1)\mu(\overline{C_s})$$

and therefore

$$\sum_{|C_s|\geq 2}|C_s|\mu(C_s) \leq 2\sum_{i,j} p_{ij}\frac{\partial\Phi_M}{\partial x_{ij}}(p_{11},\dots,p_{rr}) .$$

Consequently the series $\sum|C_s|\mu(C_s)$ diverges and the first premise of the Borel-Cantelli-lemma is performed.

To check the second premise of this lemma we show that the inequality

$$\mu(G_k\cap G_l) \quad \Big(\frac{\max p_{ij}}{\min p_i}\Big)^2\cdot\mu(G_{l-k})\cdot\mu(G_k) \tag{10}$$

holds. Using $G_k = \bigcup_{|C_s|\geq k} T^k\overline{C_s}$, $G_l = \bigcup_{|C_s|\geq l} T^l\overline{C_s}$, it is sufficient to estimate the measure of the intersection $T^l\overline{C_s}\cap T^k\overline{C_t}$. From corollary of Lemma 1 follows that the cylinder $T^l\overline{C_s}$ intersects the cylinder $T^k\overline{C_t}$ iff C_s is a subword of the word C_t , i.e. $C_t = V_1C_sV_2$, where V_1 is the beginning of the word C_t , V_2 its end and $|V_1| = k-l$.

Therefore

$$\mu(T^l\overline{C_s}\cap T^k\overline{C_t}) \leq \Big(\frac{\max p_{ij}}{\min p_i}\Big)^2\cdot\mu(\overline{C_s})\cdot\mu(\overline{V_1V_2}) .$$

Because of $V_1V_2 \in \mathcal{E}(K)$ and because this word is simple, it holds

$$\mu(T^{1}\overline{C}_s \cap G_k) \leq \left(\frac{\max p_{ij}}{\min p_i} \right)^2 \cdot \mu(\overline{C}_s)\cdot \mu(G_{k-1}) .$$

We obtain (10) by summing-up the last inequalities according s. From inequalities (10) we have

$$\sum_{i,j=1}^{n} \mu(G_i \cap G_j) \leq \sum_{i=1}^{n} \mu(G_i) + 2\left(\frac{\max p_{ij}}{\min p_i} \right)^2 \sum_{i,j=1}^{n} \mu(G_i)\cdot\mu(G_j) .$$

Therefore by divergence of the series $\sum \mu(G_k)$ follows that

$$\lim_{n\to\infty} \left[\sum_{j=1}^{n} \mu(G_j)\right]^{-2} \sum_{i,j=1}^{n} \mu(G_i \cap G_j) \leq \lim_{n\to\infty} \left[\sum_{j=1}^{n} \mu(G_j) \right]^{-1} +$$

$$+ 2\left(\frac{\max p_{ij}}{\min p_i} \right)^2 \lim_{n\to\infty} \frac{\sum_{i,j=1}^{n} \mu(G_i)\cdot\mu(G_j)}{\left[\sum_{j=1}^{n} \mu(G_j) \right]^2} = 2\left(\frac{\max p_{ij}}{\min p_i} \right)^2 > 0 .$$

Because the set Ω_K' is shift-invariant and its measure is positive, it holds $\mu(\Omega_K') = 1$. So the proof is concluded.

<u>Example</u>. Let $A = \{a_1, a_2\}$, $K = \{a_1, a_2\}$. We will find conditions by which the code K covers the Markov measure μ with transition matrix

$$\Pi = \begin{pmatrix} p & 1-p \\ 1-q & q \end{pmatrix} .$$

To that aim we make use of the criterion just proved. In accordance with the example to Theorem 3 the generating function $\Phi_M^{1,2}$ is solution of the equality

$$x_{11}x_{21}x_{22}\Phi^2 + (x_{11}x_{22} + x_{21}x_{12} - 1)\Phi + x_{12} = 0 .$$

The point (p , 1-p , 1-q , q) is singular for the function $\Phi_M^{1,2}$ if and only if at this point the discriminant

$$D = 1 + x_{11}^2x_{22}^2 + x_{21}^2x_{12}^2 - 2x_{11}x_{22} - 2x_{21}x_{12} - 2x_{11}x_{21}x_{12}x_{22}$$

is equal to zero.

It is $\quad D(p , 1-p , 1-q , q) = (p - q)^2$.

Therefore $D = 0$, if $p = q$. Direct check shows that the condition

$$\sum p_{ij} \frac{\partial \Phi_M^{1,2}}{\partial x_{ij}} (p_{11}, p_{12}, p_{21}, p_{22}) = \infty$$

is fulfilled for points of the form (p , 1-p , 1-q , q) .

So the code we consider covers measure μ if and only if the matrix π is symmetric. But this corresponds to the intuitive idea (basing upon ergodic theorem) of the same number of transitions $a_1 \longrightarrow a_2$, $a_2 \longrightarrow a_1$ in the typical words of $\mathcal{E}(K)$.

For Bernoulli measures μ the necessary condition for (connectedly) covering without overlapping by the code has the following form:

a) the onedimensional distribution $(p_1, p_2, \ldots, p_r)$ is a singular point of the function Φ_B

and b) $\sum p_i \frac{\partial \Phi_B}{\partial x_i}(p_1, \ldots, p_r) =$.

The condition that the code K in Theorem 4 is a code without overlapping is essential. Indeed, let us consider the code $K = \{a_1 a_2 a_1\}$ in the alphabet $A = \{a_1, a_2\}$. Following the proof of Theorem 2 one can show that in this case the generating function $\Phi_B(x_1, x_2)$ satisfies the equation

$$1 - \Phi_B(x_1, x_2) + x_1^2 x_2 \Phi_B^3(x_1, x_2) = 0 .$$

The point $(\frac{2}{3}, \frac{1}{3})$ is singular for the function Φ_B and the series

$$\frac{2}{3} \frac{\partial \Phi_B}{\partial x_1} + \frac{1}{3} \frac{\partial \Phi_B}{\partial x_2}$$

diverges at this point. For the property to be a word in $\mathcal{E}(K)$ it is necessary that this word doesn't contain the subword $a_2 a_2$. But the measure of the set of all sequences containing the subword $a_2 a_2$ is positive. Therefore the considered code doesn't cover a Bernoulli measure.

In some cases Theorem 1 allows a conversion for codes with overlapping.

<u>Proposition 2</u>. If K is a code such that no simple word is the beginning or the end of another simple word and the series Φ_B diverges at the point $(p_1, \ldots, p_r)$, then the code K covers Bernoulli measure with one-dimensional distribution disconnectedly.

Proof. From the condition and from relation (6) it follows that $\Psi_B(p_1, \ldots, p_r) = 1$. Because the cylinders $\overline{C_j}$ don't intersect we have

$\sum p(\overline{C_j}) = 1$. Therefore $\{\overline{C_j}\}$ is a partition. Let $\omega \in \overline{C_{i_1}}$, $T^{|C_{i_1}|}\omega \in \overline{C_{i_2}}$, $T^{|C_{i_2}|}\omega \in \overline{C_{i_3}}$, ... Then the sequence of symbols $\omega_0\omega_1\omega_2\ldots$ has a representation in the following product form $C_{i_1}C_{i_2}C_{i_3}\ldots$.

We use the notation C_j^* for the word we construct from C_j by inversion of the succession of its letters and we denote by ω^* the sequence $TR\omega$ where $R\omega = \{\ldots, \omega_1, \omega_0, \omega_{-1}, \ldots\}$. The cylinders $\overline{C_j}^*$ don't intersect, too, and it holds $\sum p(\overline{C_j}^*) = 1$. If $\omega^* \in C_{j_1}^*$, $T^{|C_{j_1}|}\omega^* \in \overline{C_{j_2}}^*$, ... , then the sequence $\omega_{-1}\omega_{-2}\ldots$ can be represented in the product form $C_{j_1}^*C_{j_2}^*\ldots$.

So we proved that a.s. the sequence $\omega = \{\ldots, \omega_{-1}, \omega_0, \omega_1, \ldots\}$ has a representation in form of a twin-countable product of simple words. The proof is finished.

Example. Let $A = \{a_1, a_2\}$ and K the code consisting of the words a_1 , a_2^3 . The words of the form a_1 , a_2^3 , $a_2a_1a_2^2$, ... , $a_2a_1^ka_2a_1^la_2$ form the class of all simple words from $\mathcal{E}(K)$. Therefore

$$\Psi_B(x_1,x_2) = x_1 + x_2^3 + 2x_1x_2^3 + 3x_1^2x_2^3 + \ldots = x_1 + \frac{x_2^3}{(1-x_1)^2}$$

holds.

Obvious that no simple word is beginning or end of another simple word. Because of $\Psi_B(p,q) = 1$ for $p+q=1$ the code above covers every Bernoulli measure.

Remark. Comparing theorems 1,2 and 4 we have the result of article [7]: An S-code K covers Bernoulli measure μ if and only if the power series

$$1 - t + \sum_{W \in K} \mu(\overline{W})t^{|W|} \tag{11}$$

has multiple positive zero at the convergence circle.

Therefore the counting function Φ_B of the code K gives an explicit representation of the smallest positive solution of the equation

$$1 - t + \sum_{W \in K} \mu(W)t^{|W|} = 0.$$

For the canonical code this equation has the form

$$1 - t + \sum p_j p_o^{l_j} t^{l_j+1} = 0 \qquad (12)$$

where $\sum p_j = 1$. The smallest positive zero of equation (12) is $\frac{1}{p_o}$. This zero is multiple if $p_o = \sum l_j p_j$.

5. Variational principle

Let K be any code in the alphabet A , $\mathcal{E}(K) = \{B_j\}$, $\mathcal{E}_n(K) = \{B \in \mathcal{E}(K) : |W| \leq n\}$.

$$\alpha_K = \overline{\lim_{n\to\infty}} [\text{card } \mathcal{E}_n(K)]^{\frac{1}{n}}$$

is called the increase index of the code K.

There is the following variational characterization of the increase index of the code.

Theorem 5. $\log \alpha_K = \sup h_\mu(T)$

where supremum is over all stationary probability measures covered by the code K.

Proof. 1) First we prove that $h_\mu(T) \leq \log \alpha_K$ if the measure μ is covered by the code K and is ergodic. Let $\omega = \{\ldots, \omega_{-1}, \omega_o, \omega_1, \ldots\}$. From the Shannon-MacMillan-Breiman-Theorem it follows that

$$\frac{1}{n-m} \log \mu[\overline{\omega_m \ldots \omega_o \ldots \omega_n}] \longrightarrow h_\mu(T) \quad \text{a.e.} \qquad (13)$$

for $n-m \to \infty$. We use the notation

$$\Delta = \bigcap_{n=1}^{\infty} \bigcup \{T^s\overline{B}: B \in \mathcal{E}(K), |B| \geq n, 1 \leq s \leq |B|\} .$$

Let Δ' be the set of all $\omega \in \Delta$ for which relation (13) is true. For any $\theta > 0$ we define the sets $\mathcal{E}_\theta(K)$ and Δ_θ by

$$\mathcal{E}_\theta(K) = \{B \in \mathcal{E}(K): \left|\frac{\log \mu(\overline{B})}{|B|} - h_\mu(T)\right| < \theta\} ,$$

$$\Delta_\theta = \bigcap_{n=1}^{\infty} \bigcup_{n=1}^{\infty} \bigcup \{T^s\overline{B}: B \in \mathcal{E}_\theta(K), |B| \geq n, 1 \leq s \leq |B|\} .$$

We will prove that $\mu(\Delta_\theta) = 1$ for $\theta > 0$. To show this we verify the relation $\Delta' \subset \Delta_\theta$. The elements of Δ' are those sequences ω for which (13) is true as well as, moreover, the property:

for any numbers M and N there exist numbers $m < M$ and $n > N$ such that $\omega_m \ldots \omega_o \ldots \omega_n \in \mathcal{E}(K)$.

Consequently if $\omega \in \Delta'$ then for any $\theta > 0$ and numbers M,N there exist

$m < M$, $n > N$ such that $\omega_m \dots \omega_n \in \mathcal{E}(K)$ and

$$\left| \frac{1}{n-m} \log \mu[\overline{\omega_m \dots \omega_n}] - h_\mu(T) \right| < \theta .$$

Therefore $\omega \in \bigcup \{ T^s \overline{B}: B \in \mathcal{E}(K), |B| \geq n, 1 \leq s \leq |B| \}$ for $n=1,2,\dots$ and so the relation $\Delta' \subset \Delta_\theta$ is true. Because the code K covers the measure μ we have $\mu(\Delta_\theta) \geq \mu(\Delta') = \mu(\Delta) = 1$. From the Borel-Cantelli-Lemma it follows that the equality $\mu(\Delta_\theta) = 1$ holds only if

$$\sum_{B \in \mathcal{E}_\theta(K)} \mu(\bigcup_{s=1}^{|B|} T^s \overline{B}) = \infty .$$

We use the notion l_n^θ for the sum $\sum_{\substack{B \in \mathcal{E}_\theta(K) \\ |B| = n}} \mu(\overline{B})$. Then

$$\sum_{B \in \mathcal{E}_\theta(K)} \mu(\bigcup_{s=1}^{|B|} T^s \overline{B}) \leq \sum_{B \in \mathcal{E}_\theta(K)} |B| \mu(\overline{B}) = \sum_{n=1}^{\infty} n \, l_n^\theta .$$

From the divergence of the series $\sum_{n=1}^{\infty} n \, l_n^\theta$ follows that $\overline{\lim_{n \to \infty}} [l_n^\theta]^{\frac{1}{n}} = 1$.

Because of $l_n^\theta \leq 2^{-n[h_\mu(T) - \theta]} \operatorname{card} \mathcal{E}_n^\theta(K)$ we have

$$\frac{1}{n} \log l_n^\theta \leq -h_\mu(T) + \theta + \frac{1}{n} \log \operatorname{card} \mathcal{E}_n^\theta(K) .$$

For $n \to \infty$ we obtain the inequality

$$h_\mu(T) - \theta \leq \log \alpha_K$$

for any $\theta > 0$. Last for $\theta \to 0$ the above estimation for ergodic measure μ follows from this inequality.

Splitting up into ergodic components and making use of integral representation of the entropy of a shift transformation by entropies of the ergodic components we obtain that the inequality $h_\mu(T) \leq \log \alpha_k$ is true in general case.

2) Now we show that the equality sign is reached. For this end we construct the following sequence of stationary probability measures on Ω_A: for every word $B \in \mathcal{E}_n(K)$ we define the periodic sequence $\omega_B = \{ \dots, B, B, B, \dots \}$ with

$$\mu_n(T^s \omega_B) = \frac{1}{n} [\operatorname{card} \mathcal{E}_n(K)]^{-1} , \quad s=0,1,\dots,n-1 ,$$

for ω_B and its shifts $T\omega_B$, ... , $T^{n-1}\omega_B$. We remark that $\omega_B \in \Delta$ for every $B \in \mathcal{E}(K)$. We choose a sequence of natural numbers n_p so that there exists the limit

$$\lim_{p\to\infty} [\ \mathrm{card}\, \mathcal{E}_{n_p}(K)\ ]^{\frac{1}{n_p}} = \alpha_K .$$

Because the space of all stationary probability measures on Ω_A is compact in the weak topology there exists a bordering measure μ for the sequence μ_{n_p}. We will show that the code K covers this measure μ and that

$$h_\mu(T) = \log \alpha_K .$$

Because weak convergence of measures is convergence of measures on open sets of the space Ω_A, from

$$\mu_{n_p}[\bigcup_{\substack{B\in\mathcal{E}(K)\\ |B|\geq n}} \bigcup_{s=1}^{|B|} T^{s}\overline{B}] = 1 \quad \text{for every } p$$

it follows that

$$\mu [\bigcup_{\substack{B\in\mathcal{E}(K)\\ |B|\geq n}} \bigcup_{s=1}^{|B|} T^{s}\overline{B}] = 1$$

and therefore $\mu(\Delta) = 1$, too. So we have shown that the code K covers measure μ. Now we will prove $h_\mu(T) = \log \alpha_K$.

Let ξ be a finite partition of the space Ω_A with atoms $\{\omega : \omega_0=i\}$ where $i=1,\dots,r$. Then

$$\lim_{p\to\infty} \frac{1}{n_p} H_{\mu_{n_p}} (\bigvee_{t=0}^{n_p-1} T^{-t}\xi) = \lim_{p\to\infty} \frac{1}{n_p}[n_p\ \mathrm{card}\ \mathcal{E}_{n_p}(K)] = \log \alpha_K .$$

We fix n_0.

For $n_p > n_0$ holds

$$\frac{1}{n_p} H_{\mu_{n_p}} (\bigvee_{t=0}^{n_p-1} T^{-t}\xi) = \frac{1}{n_p}[H_{\mu_{n_p}}(\xi) + \sum_{t=1}^{n_p-1} H_{\mu_{n_p}}(\xi | \bigvee_{i=1}^{t} T^{-i}\xi)] \leq \tag{14}$$

$$\leq \frac{1}{n_p}[H_{\mu_{n_p}}(\xi) + H_{\mu_{n_p}}(\xi|T^{-1}\xi) + \dots + H_{\mu_{n_p}}(\xi|\bigvee_{i=1}^{n_0-1}T^{-i}\xi) + (n_p-n_0)H_{\mu_{n_p}}(\xi|\bigvee_{i=1}^{n_0}T^{-i}\xi)].$$

It holds

$$H_{\mu_{n_p}}(\xi|\bigvee_{i=1}^{n_0}T^{-i}\xi) \longrightarrow H_\mu(\xi| \bigvee_{i=1}^{n_0} T^{-i}\xi) . \tag{15}$$

Because of

$$h_\mu(T) = \lim_{n_0\to\infty} H_\mu(\xi | \bigvee_{i=1}^{n_0} T^{-i}\xi) \tag{16}$$

and regarding (15) and (16), from (14) for $p\to\infty$ the inequality

$$h_\mu(T) \geq \log \alpha_K$$

follows. The proof of the theorem is concluded.

One can interpret the value log α_K in a natural way as capacity of the code K. Theorem 5 shows the following property:

If the entropy $h_\mu(T)$ of a given source $[\Omega_A , \mu]$ is greater than the capacity of the code K then doesn't exist coding defined by the code K. On the other hand this theorem says that for every code K with alphabet A there exists a source $[\Omega_A , \mu]$ which allows coding by the code K. Thereby the capacity of the code K (regarded as channel) is reached on this source.

There is also another application of the variational principle. In the paper a combinatorial characteristic of the massivity of a subgroup of a free group is introduced. The variational principle shows that this characteristic is of entropy kind.

<u>Theorem 6</u>. The maximal entropy measure for a finite S-code is uniquely determined. It is an image of a Bernoulli measure by coding by code K.

Proof. First we show that the maximal entropy measure for the canonical code $K_o = \{W_j = \underbrace{a_o \ldots a_o}_{l_j} a_j\}_{j=1}^r$ is unique and is a Bernoulli measure P with one-dimensional distribution $p_o, p_1, \ldots, p_r$,where $p_o = \frac{1}{\beta}$, $p_j = \alpha_{K_o}^{-(l_j+1)} \beta^{l_j}$, j=1,...,r , and β is a multiple positive zero of the equation

$$1 - t + \sum_{j=1}^{r} \alpha_{K_o}^{-(l_j+1)} t^{l_j} = 0 . \qquad (17)$$

The function $\mathcal{E}(t) = \sum_{n=0} \operatorname{card} \mathcal{E}_n(K) t^n$ is a solution of the equation

$$1 - \mathcal{E}(t) + \sum_{j=1}^{r} t^{|W_j|} \mathcal{E}^{|W_j|}(t) = 0 .$$

Because $\frac{1}{\alpha_{K_o}}$ is the nearest to zero singular point of the power series $\mathcal{E}(t)$, the equation (17) has the multiple positive zero $\beta = \mathcal{E}(\frac{1}{\alpha_{K_o}})$. Since β fulfils equation (17), $(p_j)_o^r$ is a probability distribution. By direct computation we obtain that the entropy of this distribution is equal to log α_{K_o} . The Lifschitz-condition [7] shows that the Bernoulli measure with one-dimensional distribution $(p_j)_o^r$ is covered by the code K_o . So this measure is maximal entropy measure of the code K_o .

Now we show the uniqueness of maximal entropy measure. Let μ be a measure covered by the code K_o and with the one-dimensional distribution $\mu_o, \mu_1, \ldots, \mu_r$. We denote by ν the Bernoulli measure with the same one-dimensional distribution. If μ doesn't be a Bernoulli measure then $h_\nu(T) > h_\mu(T)$. We prove that measure ν is covered by the code K_o. Because μ is covered by the code K_o, there exists a sequence $\omega \in \Omega$ with the following property: for any M,N exist such numbers $m < M$, $n > N$ that the word $\omega_m \ldots \omega_n$ is in $\mathcal{E}(K_o)$. Since the code K_o is canonical it follows that

$$|\omega_m \ldots \omega_n| = \sum_{j=1}^{r} l_j |\omega_m \ldots \omega_n|_j$$

for every $\omega_m \ldots \omega_n \in \mathcal{E}(K_o)$. Regarding the limit for $n-m \rightarrow \infty$ in the equation

$$\frac{1}{n-m} |\omega_m \ldots \omega_n|_o = \frac{1}{n-m} \sum_{j=1}^{r} l_j \, |\omega_m \ldots \omega_n|$$

we obtain (in consequence of ergodic theorem) a necessary condition for covering:

$$\mu_o = \sum_{j=1}^{r} l_j \mu_j \ .$$

Well known [7] that for Bernoulli measures this condition is sufficient, too. Therefore measure ν is covered by code K_o . So we have shown that the maximal entropy measure for the canonical code is Bernoulli. It is to remark that the one-dimensional distribution $p_o = \frac{1}{\beta_o}$, $p_j = \alpha_{K_o}^{-(l_j+1)} \beta^{l_j}$, $j=1,\ldots,r$, (where β is a multiple positive zero of the equation (17)) is the unique maximal point of the function $-\sum p_j \log p_j$ under condition $p_o = \sum l_j p_j$.

The case of any S-code K you like one reduces to the above by coding $\varphi\colon K \longrightarrow K_o$, where K_o is the canonical code to code K. We remark that for codes with overlapping maximal entropy measure in general doesn't be unique.

<u>Example</u>. We regard alphabet $A = \{a_o, \ldots, a_4\}$ and code $K_o = \{a_o a_1, \ldots, a_o a_4\}$.

The generated function $\mathcal{E}(t)$ of the code K_o fulfils the equality

$$1 - \mathcal{E}(t) + 4t^2 \mathcal{E}^2(t) = 0 .$$

From this it follows that the increase index α_{K_o} is equal to 4 and $\beta = \mathcal{E}(\frac{1}{4}) = 2$. Therefore the maximal entropy measure of the code K_o is Bernoulli and has the distribution

$$p_o = \frac{1}{2} , \ p_1 = p_2 = p_3 = p_4 = \frac{1}{8} .$$

Since the code K_o is canonical for the code $K=\{b_1b_3, b_1b_4, b_2b_3, b_2b_4\}$ the maximal entropy measure of code K is also Bernoulli with one-dimensional distribution $q_1 = q_2 = q_3 = q_4 = \frac{1}{4}$.

References.

[1] Clifford, A.H. and G.B.Preston: The algebraic theory of semigroups, v.1 . Providence, Rhode Island 1961.

[2] Grigorchuk, R.I.: Random walks on discrete groups. Multicomponent random systems. Adv. in prob. and rel. topics 6(1980), 285-325.

[3] Greenleaf, F.P.: Invariant means on topological groups and their applications. Van Nostrand Math. Studies 16(1969).

[4] Saslavskij, A.Ch.: On the isomorphism problem for stationary processes. Teorija Verojatn. i ee prim. 9/2(1964), 318-326.

[5] Keane, M. and M. Smorodinsky: The finitary isomorphism theorem for Markov shifts. Bull.Amer.Math.Soc. 1/2(1979), 436-438.

[6] Levenschtein, V.I.: Decoding machines, invariant with respect to the initial state. Sb. Probl. Cyb. 12(1964), 125-136.

[7] Livschitz, A.N.: On the isomorphism problem for Bernoulli schemes. Teorija Verojatn. i ee prim. 19/2(1974), 409-416.

[8] Magnus, W., A. Karrass and D. Solitar: Combinatorial group theory. New York 1966.

[9] Meshalkin, L.D.: One example for isomorphism between Bernoulli schemes. Dokl. AN SSSR 128/1(1959), 41-44.

[10] Spitzer, F.: Principles of random walk. Princeton, New Jersey 1964.

[11] Vere-Jones, D.: Ergodic properties of non-negative matrices, II. Pacific J. Math. 26/3(1968), 601-620.

ADDRESSES OF PARTICIPANTS OF THE CONFERENCE

Dr. sc. Alberti, P.M.,	KMU Leipzig , Sektion Physik , 7010 Leipzig , Karl-Marx-Platz 10 , DDR
Böhme, O.,	TU Dresden , Sektion Mathematik 8045 Dresden , Lilienthalstr. 12 , DDR
Dr. Crell, B.,	KMU Leipzig , Sektion Physik , 7010 Leipzig , Karl-Marx-Platz 10 , DDR
Dr. Der, R.,	Zentralinstitut für Isotopen- und Strahlenforschung der AdW der DDR , 7050 Leipzig , Permoserstr. 15 , DDR
Dr. Dravecký, J.,	Katedra matematickej analýzy MFF UK , 816 31 Bratislave , Mlynská dolina , CSSR
Franke, M.,	Koordinierungsstelle für Feldversuchswesen der AdL , 4204 Bad Lauchstädt , Hallesche Str. 44 , DDR
Jentsch, Th.,	MLU Halle-Wittenberg , Sektion Mathematik , 4010 Halle , Univ.-Platz 6 , DDR
Mathiszik, B.,	MLU Halle-Wittenberg , Sektion Mathematik , 4010 Halle , Univ.-Platz 6 , DDR
Dr. Naumann, W.,	Zentralinstitut für Isotopen- und Strahlenforschung der AdW der DDR , 7050 Leipzig , Permoserstr. 15 , DDR
Prof. Dr. Szlenk, W.,	Institute of Applied Mathematics and Statistics , S66W - Agricultural Univ. , Warsaw , ul. Nowoursynowska 166 , Poland
Warstatt, V.,	MLU Halle-Wittenberg , Sektion Mathematik , 4010 Halle , Univ.-Platz 6 , DDR
Dr. Weihert, E.,	MLU Halle-Wittenberg , Sektion Mathematik , 401 Halle , Univ.-Platz 6 , DDR

Dr. Wobst, R., Zentralinstitut für Festkörperphysik und Werkstofforschung der AdW , 8027 Dresden , Postfach , DDR

Zylka, Ch., KMU Leipzig , Sektion Physik , 7010 Leipzig , Karl-Marx-Platz 10 , DDR

Further addresses may be found on the first page of corresponding articles in this volume.

Bei Fragen zur Produktsicherheit wenden Sie sich bitte an:
If you have any questions regarding product safety,
please contact:

Walter de Gruyter GmbH
Genthiner Straße 13
10785 Berlin
productsafety@degruyterbrill.com